Animal re-introductions: the Arabian oryx in Oman

Cambridge Studies in Applied Ecology and Resource Management

The rationale underlying much recent ecological research has been the necessity to understand the dynamics of species and ecosystems in order to predict and minimise the possible consequences of human activities. As the social and economic pressures for development rise, such studies becomes increasingly relevant, and ecological considerations have come to play a more important role in the management of natural resources. The objective of this series is to demonstrate how ecological research should be applied in the formation of rational management programmes for natural resources, particularly where social, economic or conservation issues are involved. The subject matter will range from single species where conservation or commercial considerations are important to whole ecosystems where massive perturbations such as hydro-electric schemes or changes in land-use are proposed. The prime criterion for inclusion will be the relevance of the ecological research to elucidate specific, clearly defined management problems, particularly where development programmes generate problems of incompatibility between conservation and commercial interests.

ANIMAL RE-INTRODUCTIONS

the Arabian oryx in Oman

Mark R. Stanley Price

Office of the Adviser for Conservation of the Environment
Diwan of Royal Court
Muscat, Sultanate of Oman

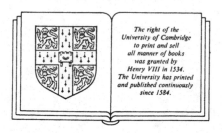

The right of the
University of Cambridge
to print and sell
all manner of books
was granted by
Henry VIII in 1534.
The University has printed
and published continuously
since 1584.

CAMBRIDGE UNIVERSITY PRESS

Cambridge

New York   Port Chester

Melbourne   Sydney

CAMBRIDGE UNIVERSITY PRESS
Cambridge, New York, Melbourne, Madrid, Cape Town, Singapore,
São Paulo, Delhi, Dubai, Tokyo

Cambridge University Press
The Edinburgh Building, Cambridge CB2 8RU, UK

Published in the United States of America by Cambridge University Press, New York

www.cambridge.org
Information on this title: www.cambridge.org/9780521131674

First published 1989
This digitally printed version 2010

A catalogue record for this publication is available from the British Library

Library of Congress Cataloguing in Publication data

Stanley Price, Mark R.
Animal re-introductions.
(Cambridge studies in applied ecology and resource management)
Includes index.
1. Arabian oryx – Oman. 2. Animal introduction – Oman.
3. Mammals – Oman. I. Title. II. Series.
QL737.U53P75   1989   639.9   88-25810

ISBN 978-0-521-34411-1 Hardback
ISBN 978-0-521-13167-4 Paperback

This book is dedicated to

His Majesty Sultan Qaboos bin Said,

whose faith in the possibility of returning the oryx to his country
has supported the project throughout, with admiration for his
appreciation that the conservation of wildlife promotes
development through the special skills and interests of local
people

The Publishers gratefully acknowledge the collaboration and financial assistance of the Office of the Adviser for Conservation of the Environment, Diwan of Royal Court, The Palace, Muscat, Oman.

CONTENTS

Contents ix

The colour plates are situated between pp. 28 and 29*
*These plates are available for download in colour from
www.cambridge.org/9780521131674

FOREWORD

It is a very special pleasure for me to contribute a foreword to this book, because I have enormous admiration for its author and the success of his project in Oman.

For some years, the Arabian oryx symbolised man's shameful destruction of the natural world around him. He exterminated from the wild a beautiful and distinctive antelope which had evolved to exploit one of the world's harshest environments. Its ultimate restoration to the deserts of Arabia is now a very real possibility, as this book shows, with its account of the planning, problems and successes of the first 7 years of the oryx re-introduction project in Oman.

I hope that this story is the beginning of the final stages of a classic conservation success story which began with Operation Oryx and the capture of a few of the last wild oryx in 1962, their increase in secure captivity in the USA, and the return of their descendants to Arabia. I have been actively involved with this sequence at all stages as Chairman and then President of the Fauna and Flora Preservation Society, which sponsored the original expedition, and as Chairman of the International Union for the Conservation of Nature Species Survival Commission.

It is particularly encouraging that the initiative to re-establish the oryx came from the head of state of the Sultanate of Oman. His Majesty Sultan Qaboos bin Said has demonstrated his interest in wildlife and his commitment to a productive natural environment for the lasting benefit of his people, through many projects. These include the protection of nesting turtles and their eggs, a mountain reserve for the endemic Arabian tahr and the establishment of a breeding centre for them and other endangered Omani species. Systematic surveys of the Sultanate's fauna and flora have been made to document the rich resources of this desert country, which lies at the junction of three major zoogeographical realms.

I have visited Oman several times between 1976 and 1987. In 1983, my wife and I flew into the oryx re-introduction area, landing by helicopter deep in the desert a few kilometres from the first herd of wild oryx and the assembled ranger force. In the immensity of the desert the herd seemed a frail resource, but it would clearly not suffer from lack of care and protection by the local people. The oryx were a dramatic sight, with several desert-born calves, as they moved around in command of their ancestral landscape, and now once more an aesthetically integral part of it.

Since then, further oryx have been released, the herds have ranged further, and new insights into re-introduction methods deduced. Oman's oryx project is an example to others of what can be done with the right will and dedication, and with support from the local community.

As so many of the world's ecosystems, especially the deserts, are increasingly degraded and denuded of their wildlife, Oman is to be congratulated on showing that this trend can be reversed. It is a pleasure to introduce Mark Stanley Price's book, and I hope that the story it tells will inform and entertain everyone interested in contemporary conservation.

Peter Scott

PREFACE AND ACKNOWLEDGEMENTS

The Arabian or white oryx is a charismatic animal. It features frequently in poetry of the Arabian peninsula, with many references to the beauty of its eyes. Known to the ancient Greeks as a captive animal in Egypt, Aristotle recorded the habit of binding together the growing horns of young oryx so that they fused into one. Thus, the unicorn entered mythology from a firm factual basis. In the western world the unicorn, in remarkable contrast to other beasts of the imagination, was not held in fear but was projected as noble, chaste, fierce yet beneficent, altruistic though solitary, and strangely beautiful (Shepard, 1930). Unfortunately, these attributes did not confer immortality, and over hundreds of years the unicorn was pursued and hunted progressively from its furthest desert fastnesses. The last herd of wild Arabian oryx was destroyed in 1972.

This book is the story of the oryx's return from captivity to the desert in the Sultanate of Oman; why Oman wanted the oryx back and was able to offer it a secure home. The local people of the re-introduction area are central to the story. They are the Harasis (singular, Harsusi), belonging to the large body of desert tribes whose nomadic, pastoralist life-style makes them *bedu*, a more correct form than *beduin*. Without attempting to be authoritative on them, I have written of the Harasis in some detail because of the relevance of their ecology and development to the oryx.

The re-introduction project was less research than a scientifically planned and monitored management exercise. The sense of pioneering in the desert encouraged us to record many observations on the oryx, the desert and its wildlife that do not appear in this book. The scientific results are presented here, but many are buttressed by anecdotal observations on the behaviour of individual animals. Apart from offering insights which no amount of systematic data collection could have feasibly matched, we knew each oryx well through its very individual appearance and character. By the end, we had known most oryx from birth, through growth and, in some cases, to their deaths.

The re-establishing oryx were few in number. As they increased, intensive observation and surveillance of them illuminated the plight of small populations of wild animals in their fight to stave off extinction. This is a very fashionable topic in contemporary conservation biology, and one where theory has run ahead of field data. Moreover, the first oryx to return to Oman were zoo-bred. Hence, the project was a testing ground for the ability of animals, several generations removed from the wild, to adapt to, and thrive in, their ancestral habitat. Fortunately for zoos and for conservation, the oryx generally acquitted themselves very well.

As animal re-introductions are now so topical – as a glance at any issue of *Oryx*, the journal of the Fauna and Flora Preservation Society will confirm – I have attempted to analyse other re-introduction efforts, and to place the oryx story in context. The review of re-introductions in Chapter 1 is eclectic, and could probably never be complete because of the obscurity of much of the literature. Some case-histories are referred to frequently, usually because they are well documented and their techniques have been analysed. With this background, the oryx's return is used in Chapter 10 to develop a general theory for re-introduction.

My wife and I spent over 7 years with the oryx in the deserts of central Oman. This book cannot convey the excitement in watching and marvelling at Harasis rangers tracking a single oryx over a stony, pavement surface, or receiving a morning radio report of a new, wild-born calf. There were anxieties too, such as sick animals, lost radio contact with patrols, or only one day's water supply in base camp. The best antidote to nostalgia is to contemplate the future, and, as I write, four new oryx from Jordan are in Oman and another six have just arrived from the USA. In due course, these immigrants will swell the wild population, which is enjoying a record crop of winter calves.

The following chapters will show that many people were involved in Oman's oryx saga, and assisted most willingly in any way they could. It is a pleasure and honour for me to thank Sir Peter Scott for his generous foreword and to acknowledge his enthusiasm, which sustained the oryx project through its early years of planning.

The project received the fullest support from the Diwan of Royal Court, especially from the Presidents, HE Sayyid Saif bin Hamed bin Saud al-Busaidi and his predecessor, HE Lt. Col. Said bin Salem al-Wahaibi. Michael Harford ensured a reliable flow of funds and a recurrent budget with minimal formality. The late Bruce Blackwell of the Diwan Properties Office cared for the camp and its machinery with great devotion. The Royal Oman Police and the Ministry of the Interior, through successive Walis and Naib Walis of Haima, assisted the oryx project whenever necessary, and I acknowledge the special co-operation of Shaikh Abdullah bin Mohamed bin Ghaith al-Darmaki.

The Office of the Adviser for Conservation of the Environment backed up the project in Muscat and handled countless problems and requests over the radio. It is a pleasure to thank Ralph Daly, Yousuf bin Hussein bin Mohamed al-Baluchi, Khamis bin Said al-Shikeli, Mahmood bin Abdul Kardar al-Shahwarzi, Rafique Ahmed and Eileen Scarff for their particular kindnesses and enthusiasm. The assistance of Faisal bin Mohammed bin Sulaiman al-Lamki was invaluable in conversations with the Harasis about their tribal past. We enjoyed the enormous hospitality of Ralph and Elizabeth Daly whenever in the capital.

In the desert, the oryx were the concern of the entire Harasis tribe, for which I am indebted to all shaikhs and tribesmen, and especially to Shaikh Ibrahim bin Sagar al-Harsusi, who was an excellent mentor and friend in the early and uncertain days. Everyone employed at Yalooni camp gave sterling service, but I thank particularly the head rangers, Said bin Dooda and Nasir bin Rashid, two remarkable Harasis who taught us much about their desert home, and Munaf Kazi and Kulwant Singh for their roles in orderly administration and keeping the generators and vehicles running.

Tim Tear, assistant manager for three years and now manager, was an outstanding helper and companion for testing new ideas and theories on oryx behaviour, and for sharing the load when 35 oryx were scattered across the desert and had to be found. It is also a pleasure to thank Mark Infield, Nick Lane Fox, Justin Bell, Angus McLeod, Geoff Gamlen and Simon West for their services as field assistants, map-makers and data-jugglers at Yalooni for short periods.

The oryx's veterinary matters were capably looked after by Dr Michael Woodford, and Drs Jeff Kidner, Ian MacLeish and Ray Ring of the Diwan, and Dr Abdul Gaber Saleh of the Ministry of Agriculture. I am most grateful to them all for their wise counsel, and tolerance of visits to Yalooni in noisy aircraft, often at very short notice. Jeremy Usher Smith of the Oman Endangered Species Breeding Centre responded nobly to crises involving abandoned animals, milk formulae and concentrations. The Liverpool School of Tropical Medicine diagnosed snake-bite as the cause of death of one oryx.

Michael Gallagher of Oman's Natural History Museum helped greatly by identifying any bird or animal found dead or alive. The Royal Botanic

Garden at Edinburgh identified all plant specimens collected, and particular thanks are due to Tony Miller, Rose King and Ian Hedge. Dr Hartmut Jungius, who carried out the project's feasibility study for the World Wildlife Fund, remained a stalwart ally and visited the project over the years.

The Sultan of Oman's Air Force brought all the oryx to Yalooni, and aircraft of various squadrons flew in and out on many tasks – medical evacuations, delivery of veterinarians, removal of dead oryx for autopsy, even flying in flood relief supplies of rations and a rubber boat. The project is most grateful for the concern and efforts of the Commander of SOAF, Air Marshall Erik Bennett WSQ, WO, CB, and all his officers for making aircraft available so promptly.

Petroleum Development (Oman) co-ordinated all aspects of its operations with the oryx project. Its Exploration Department gave information on oil installation locations, and helped with map production. Dr Michael Hughes Clarke patiently corrected my naive views on geology, landforms and hydrology. Dr Rudolf Nottrott stayed at Yalooni for a month and completed the computer program RANGE in a week less than he needed.

In 1986 we spent five months away from the desert in the fertile atmosphere of the Large Animal Research Group of the Department of Zoology at Cambridge University. The invitation was a most generous gesture by Dr Tim Clutton-Brock, and I fear that my work was the least academic ever to have gained entry there. The group could not have been more welcoming, and I am especially grateful for the efforts of Dr Iain Gordon in tracking down references in the university libraries.

In England, Dr Georgina Mace was always enthusiastic about the oryx re-introduction as a practical justification for genetic management of the captive oryx. She oriented me to the importance of inbreeding levels, genetic diversity and generation length, and spared the time to run the oryx population models.

I corresponded with many people about rumoured or ill-documented re-introduction efforts, studbook data of the 1960s and so on. Of all those who kindly responded, I would like to thank particularly Lawrence Killmar (San Diego Wild Animal Park), Maher Abu Jafer (Jordanian Royal Society for the Conservation of Nature), Michael Crotty (Los Angeles Zoo), Wayne Homan (Phoenix Zoo) and David Jones (Zoological Society of London). Michael Loyd and Michael Butler provided fascinating information about the oryx on the Jiddat-al-Harasis in the 1960s, when each travelled with camels through the area. The Fauna and Flora Preservation Society kindly allowed me to study its archives of the World Herd.

At various stages, this book has been read in its entirety by Dr Nigel

Leader-Williams on behalf of the publisher, by Ralph Daly and Tim Tear. Different chapters were also read by Maj. Ian Grimwood, Dr Michael Hughes Clarke, Dr Dawn Chatty and Dr Georgina Mace. I thank them all for their patience with rough drafts and for their suggestions, and hasten to add that any deficiencies or errors are my responsibility. I am most grateful to the African Wildlife Foundation for its encouragement for me to complete this work, and to Karen Stanley Price, who prepared most of the figures, and Deborah Snelson, who helped with the final production.

Two people contributed especially to the success of the oryx project in Oman, and this book does not do justice to them. The first is Ralph Daly, who conceived, designed and launched the oryx re-introduction and offered me the chance to manage the project. During my time in Oman he was unfailingly supportive and considerate, and a wonderful raconteur and field companion. The second is my wife, Karen, who loved Oman, the desert and its people. She contributed so much to the project through a herbarium collection, daily bird observations in an area of near total ornithological ignorance, administering first aid to all, investigating Harasis genealogy and gaining the trust and interest of the women. For all these, running a household and entertaining visitors in at least two cultures, I am more grateful than I can say.

Mark R. Stanley Price
Nairobi
March 1988

1

Re-introductions defined and reviewed

1.1   Introduction

All plant and animal species have finite life-spans on earth. However, conservationists are alarmed by the present, increasing rate of species extinctions (see, for example, Myers, 1986). The maintenance of biological diversity for natural ecosystems to the benefit of humans has become a universal creed. The simplest reason for the present extinction rate lies in the increase in numbers of human beings. The global population doubles every 40 years, but in some tropical, developing countries doubling times are as low as 18–25 years (IIED & WRI, 1987). Habitats that only 10 years ago were too distant or inhospitable are now exploited intensively, often with dire consequences for their natural resources of plants and animals.

Some plant species are preserved in botanical gardens or in germ-plasm banks, and zoos keep animals. These fortunate species tend to be those which are either striking or beautiful to the human eye, large or very rare, or useful to people. Reality intrudes with the estimated need for captive breeding over the next 200 years. Some 800 species of mammal will sidestep extinction only with this help, but the world's reputable zoos can accommodate adequate populations of less than half this number (Soulé et al., 1986).

In contrast to this bleak outlook for species diversity in the wild and limited accommodation in captivity, research is throwing up new methods and technologies for captive management and successful breeding. Around the world, also, there is a new public interest and concern for environmental issues that are seen to be linked to human welfare or, even, survival. There is renewed support for protected areas with novel and imaginative means for resource-sharing between people and wildlife. With this prospect of adequate habitat for the foreseeable future, re-introduction offers a lifeline for species endangered in the wild but with secure captive populations. It is a conservation tool with great potential for restoring communities and ecosystems to approximately their pristine states.

The Arabian oryx is an antelope which, in 1972, was exterminated from the wild, although secure in captivity. Yet, within 10 years, a few animals were again free in their desert habitat in Oman. This book describes and analyses this re-introduction effort, but, first, other re-introduction attempts, both failures and successes, are reviewed to show the multiplicity of factors involved in such an operation. (Appendix 1 contains the taxonomic names of all species mentioned.) Many of the lessons which these efforts show were heeded to the advantage of the oryx re-introduction. While the oryx developed a free-living population, many aspects of its biology were seen to promote successful re-introduction. These observations are combined in the final chapter with information from other species to assess the future potential of re-introductions. It is hoped that this book will make a small contribution to ensuring that conservation resources are used wisely on ventures with a fair chance of success.

1.1.1   *Definitions and distinctions at the population level*

There are important distinctions between introduced and re-introduced populations of animals. Introduced populations are common and include those which have become established outside their former natural range by the accidental or deliberate involvement of man. Deliberate introductions arise from biological, utilitarian or sentimental motives (Jungius, 1985), and many introductions have subsequently achieved pest status. In contrast, re-introductions are less numerous, and are invariably deliberate acts by man. Most definitions agree that a re-introduced population is living in an area of natural habitat within its former range (see, for example, Brambell, 1977; Konstant & Mittermeier, 1982; Council of Europe, 1985; Jungius, 1985; Sale, 1986). In this context, the prefix 're-' has the sense of both 'back' and 'again' (Anon., 1973), and is less restrictive than another definition: 're-introduction means an attempt to establish a species in an area to which it has been introduced but where the introduction has been unsuccessful' (Joint Committee for the Conservation of British Insects, 1986).

Some definitions of re-introduction add further specifications: the original population should have become extinct in historical times (Jungius, 1985), or the re-introduced animals should be of the available race or type closest to the aboriginal population (Greig, 1979; Sale, 1986). In some cases, these are significant criteria (1.2.4).

For animal populations, introduction and establishment, re-introduction and re-establishment have been used almost synonymously. However, (re-)-establishment implies that a viable population exists successfully, whereas (re-)introduction is only an initial stage, which may fail. Re-inforcement is the addition of animals, from whatever source, to an existing population of conspecifics.

The emphasis in all these terms is on the population rather than the individual animal. None the less, even at this level these distinctions may be blurred, as the following problem areas and examples demonstrate:

(1) *Geographical scale.* The Jamaican hutia, a medium-sized rodent, was bred in captivity and released on Jamaica in colonies (Oliver, 1985). As the island still retained natural colonies, the hutias were released in areas which had never before supported them but which were transformed into suitable habitat through the construction of artificial warrens. Genetic exchange between the two groups was unlikely because of habitat discontinuity. Thus, for the total island population this operation was not a genetic re-inforcement. If previously occupied habitat had been restocked, it would have been a re-introduction, but because of the habitat modification which resulted in range expansion, the exercise was, in fact, a local introduction.

(2) *The period since extinction and the causes of extinction.* Before the reindeer died out in Scotland in historical times (the eleventh century), it had occupied the lowland pine forests (Whitaker, 1986). Vegetation changes due to climatic change and man's activities during the next 900 years ensured that there was probably no aboriginal habitat when reindeer were released in Scotland in 1955 (Lindgren & Utsi, 1980). As they were confined to the Cairngorm alpine zone, their re-introduction to Scotland was, strictly, an introduction to a different ecological zone. This highlights the problems faced when other re-introductions are proposed to recreate the recent historical fauna of the British Isles (see, for example, Yalden, 1986).

(3) *The genetic background of animals to be re-introduced.* Some definitions of re-introduction specify that the genetic stock used should be as close as possible to the original population. The genetics of founders of a prospective re-introduced population may be critical to establishment success (1.2.4), but in some cases the original subspecies or race may no longer exist to be re-introduced. The Persian wild ass was therefore released into Israel in place of

the extinct Syrian subspecies. Strictly, this was a re-introduction of the species, but the introduction of the subspecies (Brambell, 1977). The situation is worse when another subspecies is introduced into the range of an extant local subspecies: the Canadian beaver has been introduced into the range of the living European beaver in Finland (Jenkins, 1965). Swift foxes from Colorado are to be released into the northern part of the species' range in Canada, despite the more severe climate and the persistence of a few northern foxes in the adjacent states of the USA (Stromberg & Boyce, 1986). There are many other such examples.

(4) *The effect of time.* Particularly where very mobile species are involved, a (re-)introduction may develop into a re-inforcement. Thus, the re-established sea-eagle, breeding again on the west coast of Scotland after an absence of 69 years (Love, 1983; J. A. Love, personal communication 1987), ranged far and may soon meet individuals of the western continental European populations or even of the parent Norwegian population. If genetic exchange occurs, then the re-introduction into Scotland and range extension will become a re-inforcement of the overall population. Similarly, the establishing colony of released griffon vultures in the Cévennes of France was visited by one conspecific bird from the Pyrennean population 300 km distant (Fonds d'intervention pour les rapaces, 1985). The knowledge by the Spanish bird of the new colony could be the first step towards its population's re-inforcement in future. Such developments can have serious consequences if the two populations have different genetic backgrounds (1.2.4).

1.1.2   *Definitions for individual animals, and implications for success*

It is clear that, while re-introduction is a population process, success depends on the performance of individual animals. Table 1.1 shows, with examples, how within the categories of introduction, re-introduction and re-inforcement of populations, the animals may have had various origins and been subject to different types of management. The terms are not necessarily mutually exclusive. Translocation is the moving of wild-caught animals for release into the wild at a second site. Marooning is the release of animals into an area to which they are confined, usually because it is an offshore or river island. The third major category is the release into wild or semi-wild habitat of captive-bred animals or those born in the wild but then subject to a captive existence. Preparation for release may involve rehabilitation, a process designed to teach the skills for efficient exploitation of their future habitat, with an anticipated improvement in performance and survival. As the examples show, rehabilitation is necessary in species either with complicated learned behaviours and/or those with advanced social organisations, such as apes or large carnivores.

The background of the animals and the type of release provide an approximate indication of the likelihood of successful re-introduction. Translocation between two suitable habitats makes the least demands on the animals. Marooning of captive-bred animals provides protection from the perils associated with wandering and exploring too far from the release site. Release islands may be chosen for their natural or artificial absence of potential predators. Naive, captive-bred animals have a lower chance of survival, and this premise is the basis for rehabilitating or training before release.

Another aspect of re-introduction success or the risk and cost of failure is

Table 1.1. *Terminology and examples of animal transport by man to show the effect of source and experience of the individual animals and of their destined populations*

| Species | Ref. | Origin | Type of movement at level of: | |
			Individual	Population
White rhino	a	Wild-caught	Translocation	Introduction
Indian rhino	b	Wild-caught	Translocation	Re-introduction
Baboon	c	Wild-caught	Translocation	Re-inforcing
Woolly monkey	d	Wild-caught	Marooning	Introduction
Chimpanzee	e	Caught young	Marooning	Introduction
Spider monkey	d	Wild-caught	Marooning	Re-introduction
Hutia	f	Captive-bred	Release to wild	Introduction
Arabian oryx	g	Captive-bred	Release to wild	Re-introduction
European otter	h	Captive-bred	Release to wild	Re-inforcing
Chimpanzee	i	Captive-bred/ caught young	Rehabilitation, release to wild	Introduction with marooning
Orang-utan	j	Caught young	Rehabilitation, release to wild	Re-introduction and re-inforcing
Lion	k	Captive-bred	Rehabilitation, release to wild	Re-inforcing

References: a, Booth, Jones & Morris (1984); b, Sale & Singh (1987); c, Strum & Southwick (1986); d, Konstant & Mittermeier (1982); e, Borner (1985); f, Oliver (1985); g, this work; h, Jefferies *et al.* (1986); i, Hannah & McGrew (1989); j, Aveling & Mitchell (1980); k, Adamson (1986).

the overall conservation status of the species. There is little risk in removing individuals of an abundant species from one wild population to establish another. However, the risks surrounding re-introduction increase when each individual is a valuable member of an endangered population or even of a species' total strength. In some instances the re-introducing species may be so few in number that if a first re-introduction attempt does not succeed, replacements may not be available. This obviously leads to conservative management methods. Rarity has another implication. If the re-introducing species is abundant or well-distributed elsewhere, it is likely to be well researched, with a fund of information relevant to its re-introduction methods. As species rarity intensifies, data on it will be scantier. The Arabian oryx had reached an ultimate stage where the aboriginal population was exterminated from its total range before any systematic study had been made. In such instances, indirect methods and analogies must be used to design re-introduction methods for captive-bred animals (Chapter 3).

1.2.   Prerequisites for a re-introduction attempt and its success
1.2.1   *Procedures*
Many re-introductions have been attempted, and some have undoubtedly succeeded, without any formal planning or evaluation. However, the procedures for a systematic re-introduction effort have been clearly spelt out (Anderegg, Frey & Muller, 1983; Council of Europe, 1985; Jungius, 1985; Sale, 1986; IUCN, 1987). Most activities fall into four distinct phases (IUCN, 1987).

(1) *A feasibility study that indicates the likelihood of a successful re-introduction.* If possible, the ecology of the species in its original habitat is studied, with especial emphasis on the reasons for the extinction and whether these conditions pertain any longer, and whether adequate, suitable habitat remains. A genetically suitable source of stock must be identified, whether wild or in captivity. Non-biological aspects are involved: adequate funding for all four phases over several or many years must be assured, and the re-introduction must be explained and be made acceptable to the people of the release area. This last factor is critical in the case of large and/or dangerous animals.

   If at this stage a re-introduction is not recommended, scarce conservation resources will have been saved. For example, a feasibility study examined the possibility of reinforcing a remnant, endemic skink population on Round Island, Mauritius, and concluded that re-introduction of captive-bred skinks was still inappropriate because of the high numbers of introduced rats which were the cause of the original skink population's decline (Bloxam, 1982). An effort to re-introduce the great bustard to Scandinavia was abandoned at this stage because, although technically feasible, the cost was deemed prohibitive, and there were too many potential causes of failure (Hallander, 1976).

(2) *The preparation phase.* In this phase, a release area must be identified. The sources of animals must be established, whether from wild or captive populations. Management methods appropriate to the animals' origins and experience and the climate to which they will be exposed must be designed. The numbers of founding animals, their sex and age composition must be estimated and agreement reached on their availability. Logistic arrangements for capture and transportation must be made, and the stringent veterinary requirements now usually necessary for international movement of animals must be met.

(3) *The release phase.* Planning the release includes activities such as assessing when the founding animals can be released with a high expectation of surviving. The time of day of release may be important, while selection of the correct season may be critical (1.3.6).

(4) *A post-release monitoring phase.* In the short term, monitoring the range or dispersal of released animals allows a measure of their response to the new environment. If causes of loss or death can be identified, management methods can be changed at once or for subsequent releases. Data on the animals' behaviour will identify any adverse effects on the ecosystem, in which case the re-introduction could be halted. They allow comparison of the species' biology with that of its aboriginal predecessors, as a measure of the success of the re-introduction.

While most re-introduction efforts state the value of this phase, it is often neglected once the released animals appear to be surviving adequately. Post-release monitoring of the Hawaiian goose was recommended during project design (Kear & Berger, 1980), but little took place. After the release of 1244 birds over 16 years on Hawaii and a further 391 on Maui Island, the status of the re-introduced population was not known, and the causes of their limited success at establishment could still only be surmised (Berger, 1978).

1.2.2   *Habitat availability*

It is axiomatic that no re-introduction should proceed without adequate habitat for the species. However, identification of suitable habitat may prove difficult, particularly in the case of invertebrates with very specific needs. For example, two butterflies, the large copper (Duffey, 1977) and the swallowtail (Dempster & Hall, 1980) were each released into an English fenland habitat. Both reserves had suffered from drainage of the surrounding farmlands, which led to vegetation changes that adversely affected the larval stages. Some reserve management activities were also detrimental. Under prevailing conditions it appeared that neither insect could re-establish a genuinely viable population.

The habitat occupied by the last remaining aboriginal, wild populations may be no guide to the optimum habitat for a re-introduction attempt. The last wild Przewalski horses lived in the Gobi Altai of Mongolia, where they were most remote from humans, predators and competition. However, they had a huge range throughout Europe and Asia in historic times (Ryder & Wedermeyer, 1982). Prime Père David's deer habitat was lost to agriculture in China long ago through being fertile swampland. The deer only persisted in land which flooded regularly and was difficult to drain. Thus, their re-introduction has had to be into this habitat type, which, due to its infestation with exotic species of *Imperata* grass, requires rehabilitation (Jungius & Loudon, 1985). The red wolf is being re-introduced into the coastal prairie marshes of North Carolina, not because this was as good habitat as the original bottomland riverine forests but because it provides the thick vegetation cover essential for the wolf (Parker, 1986). One of the major factors in the successful re-inforcement of the native population of Lord Howe Island woodhens through captive breeding came through the discovery that egg production by the remaining wild birds in montane habitats was many times lower than in the same birds taken to low altitudes and given supplementary feed (Fullagar, 1985).

Identification of suitable habitat is often intertwined with the presence of other exotic species. Failure by the Hawaiian goose to increase after re-introduction may have been due to its occupancy of only the higher altitudes of its historical range. Breeding formerly occurred at lower altitudes which were subsequently infested with introduced mosquitoes, vectors of avian pox to which the goose is very susceptible (Kear, 1977). The first captive-bred pink pigeons to return to Mauritius were not released to reinforce the remaining, few, wild birds. With the aim of establishing an easily monitored semi-wild population, the release area was 35 km from the wild birds. It was a Botanic Garden where almost all the plant species were exotic, and where the activities of introduced nest-predators, the Indian house-crow and Indian mynah, were a major concern (Todd, 1984).

The effects on a mammal of novel factors in the new habitat are shown in a rare case in which marooning was reversed. In a closely monitored translocation, black-tailed deer were removed from an offshore Californian island because of overpopulation, and released into mainland deer habitat with a low population of the same species. Survival amongst the released deer was low because the deer were naive in the face of hunters, predators, poachers and cars in the new habitat (O'Bryan & McCullough, 1985).

When alternative suitable habitats are available for a re-introduction, other considerations can influence the choice. Thus, the European lynx was re-introduced into large state forests in which the return of a predator would be of least concern to the public and the deer-hunting fraternity (Wotschikowsky, 1981). Conversely, the release site of golden lion tamarins was chosen to permit large numbers of the public to visit, for the tamarin became the symbol for the preservation of the natural forest itself (Kleiman *et al.*, 1986).

Decisions about habitat suitability are invariably made by humans, whereas the ultimate test is whether the re-introduced animals will thrive in it. This paradox lies behind the idea of pilot releases of surplus males of mammal species as habitat testers (Jungius, 1978*a*). However, if females are absent or atypical social groups are released (1.3.2), habitat use by such animals might not indicate its suitability for adequately prepared and natural groups.

1.2.3   *Niches and life-cycles*

Even if habitat conditions encourage a re-introduction, the re-establishing species must find its niche vacant, or at least be able to fill a niche similar to its aboriginal one. Theoretically, if a community loses a single species, the population dynamics of other species, of both the same and other trophic levels, may be expected to change and compensate. The anchoveta fish population of the Peruvian current has been severely overfished. As its role in the food web has now been filled by herbivorous crustacean plankton, any re-introduction is likely to fail (Slobodkin, 1986). No habitat or community can realistically remain unchanged in anticipation of the restoration of a missing species. Changes in the balance of species may have profound effects for re-establishing invertebrate species, and the large blue butterfly in England is a particularly well-documented case which shows the complexity of community structure influencing re-introduction success. After hatching on thyme plants, older larvae spend nine months in red ant nests, feeding on grubs. Two species of red ant co-occur, *Myrmica sabaleti* and *M. scabrinodis*, predominating under heavy or light grazing conditions, respectively. However, larval mortality is five times greater in the nests of *M. scabrinodis*, and a butterfly colony cannot achieve viability in a sward dominated by this ant. The correct grazing pressure, therefore, must be developed before any re-introduction, although thyme thrives under the range of conditions suitable to both ant species (Thomas, 1980). Marine organisms may also have specific but critical needs: culture of giant clams for restocking denuded reefs is technically possible apart from problems with infecting the larval clams with the symbiotic zooxanthellae whose presence is obligatory for normal growth (Heslinga, 1979).

Instances in which the re-introduction of vertebrates has proved a failure, or likely to be so, because the returning species has lost its niche, are rare. However, the native polecat, *Mustela putorius*, originally declined throughout Britain because of fen drainage and persecution, but its re-establishment would probably fail over large areas because of the presence of the exotic and more aggressive congener, the mink, *M. vison* (Yalden, 1986). One of the few attempts to recreate a community of large mammals shows the possible effects of an enlarged niche for a carnivore through the absence of other species. Pilanesburg National Park, in southern Africa, was restocked with a range of ungulate species and with cheetah, but without lion or hyaena, both of which compete for prey with cheetah and displace the latter from their kill. Consequently, the cheetah population increased to such an extent that they eliminated the re-establishing population of springbok, and had to be reduced in numbers (Anderson, 1986).

1.2.4 *Genetic suitability of founders*

Because animals adapt genetically to prevailing conditions, whether in captivity or in the wild, founders for re-introduction must be as similar as possible genetically to the aboriginal population. The greater the distance between source and release areas, the less likely are the chances of successful establishment due to genetic maladaptation. This may be particularly important in the case of terrestrial invertebrates with highly controlled life-cycle stages. For example, the founder stock of large blue butterflies for re-introduction into England could not come from adjacent areas of continental Europe because their strains were also extinct (Jones, 1986). In southern Europe the butterfly lays its eggs, not on thyme as in England, but on marjoram which would result in too late an emergence in England. Consequently, butterflies were taken from the Swedish strain, which also feeds on thyme. Although thyme flowers in July in Sweden, butterflies of

Swedish origin hatch in England in June, in synchrony with the local thyme (Jones, 1986).

It is hard to document any re-introduction failure that was due solely to genetic maladaptation. However, difficulties in re-establishing the great bustard in England may have been due to the founders coming from Portugal and Hungary (Osborne, 1985). The climate of each country differs, with implications for the chronology of nesting in relation to season and insect food abundance. The deleterious effects of inappropriate genetic composition are also often shown where remnant populations are re-inforced by stock of very different origin. In an effort to 'strengthen' a wild turkey population, domestic turkeys were released in the USA (Mayr, 1963). Contrary to the expected result, hybrid birds had a far lower breeding success. Even more dramatic were the results from successive introductions of the ibex, *Capra ibex*, into Czechoslovakia (Turček, 1951). After the original population of the Tatra mountains was hunted to extinction, it was replaced by *C. ibex* from Austria. A few years later it was reinforced by *C. hircus aegagrus* from Turkey and *C. ibex nubiana* from Sinai. The resulting three-way hybrids dropped their kids three months earlier than the pure *C. ibex* population, in the coldest season, resulting in total kid loss. Hence, extinction of the population was most likely to have been due to the genetically determined breeding season suitable for lower latitudes (Greig, 1979).

In some circumstances a re-introduction may only be possible with founder stock from areas distant from the re-establishment site. The highly successful re-introduction of the peregrine falcon into the eastern USA was achieved by 'hacking' (1.3.5) into the wild young birds derived from the genetic stocks of Spain, Scotland, Chile, Alaska, the Aleutian Islands, the Queen Charlotte Islands and California. Despite criticism that these birds must have been genetically very different from the aboriginal population, there were no more suitable sources, and the environment had, in any event, changed since the extinction, so that the re-establishing birds had to face a new set of selective forces. Whether or not the re-established population converges in appearance to the original birds or differs slightly in response to modern selective forces makes little difference to most peregrine watchers (Barclay & Cade, 1983). It may also be unrealistic to take too purist an approach to the genetic background of founders for re-introduction in the case of captive-bred animals. Often, subspecies cross to an unknown degree and in unknown combinations, as in the case of hundreds of eagle owls released into Germany, which are now successfully re-established (von Frankenberg, Herrlinger & Bergerhausen, 1983). It was felt that captive-breeding irrespective of subspecies probably resulted in a fairly uniform genotype in the released birds.

1.3   Management techniques for re-introduction

The preparation phases of a re-introduction are needed to design the release strategy with the broad aim of establishing viable numbers of animals with as low a loss as possible in their new habitat. There are many aspects to this, such as the background and age of the individual founders and the composition of the released groups. The chances of survival can, in many cases, be enhanced by management methods directed either at habitat modification or reproductive manipulation, by the provision of supplementary feed and by restricting the dispersion of the released animals. These techniques in re-introductions are described in this section.

1.3.1   *Founder selection*

When released animals are to be exposed to the selective forces of their aboriginal environment, their survival prospects are decreased by any physical or behavioural abnormality. This is particularly important where captive-bred individuals are the re-introduction stock. For example, in one zoo, 50% of the bearded vultures, potential candidates for release in central Europe, were unsuitable because of physical damage, imprinting on humans or through other abnormal behaviour (Anderegg *et al.*, 1983). Only under very specific circumstances may animals that are habituated to people fare better on release into the wild: captive-bred beavers released in areas of human occupancy performed better than translocated, wild-bred animals (Zurowski, 1979). Genetic normality is also preferable, for high inbreeding levels are likely to predispose to post-release loss: in the first release of pink pigeons on Mauritius, one bird had an inbreeding coefficient of 0.25, and two others had inclined feet, with a possible genetic basis (Todd, 1984).

The ages of released animals can be critical to re-introduction success, but generalisations are difficult, for species-specific considerations may determine the optimum age structure. The first adult, captive-bred golden lion tamarins to be released into Brazil showed less adaptability at exploiting the complex environment, and less locomotor ability and skill at predator detection than young animals (Kleiman *et al.*, 1986). Future releases will comprise age-graded family groups for the sake of the social structure (1.3.2) and to increase the potential of young animals to adapt to their native habitat. For birds, the release of young individuals is generally recommended as they are primed to learn the necessary survival skills better than adults (Temple, 1983). Experience with chimpanzees suggests that adolescents or young adults, even if wild-born but then subject to a captive existence, respond better subsequently to wild surroundings than older animals (Borner, 1985).

The release of young individuals results in the trade-off that such animals take longer to reach breeding age. Amongst the many large birds of prey being re-introduced, the sea eagle first breeds at 5 years old (Love, 1983) and the griffon vulture at 6 years (Fonds d'intervention pour les rapaces, 1985). Birds released at fledging during their first year may, therefore, die before reaching breeding age. For this reason releases of griffon vultures of at least 6 years of age and over several consecutive years have been recommended to establish a breeding colony at a specific cliff site and to compensate for losses during establishment (Parc National des Cévennes, 1981). Similarly, breeding in translocated sea otters appeared to be delayed because those moved contained a high proportion of subadults which dispersed (Jameson *et al.*, 1982). Experience with muskox translocations has also shown that released herds containing a high proportion of subadults had low initial rates of increase, and 5–6 years passed before breeding compensated for the mortality due to the move (Grauvogel, 1984).

A successful re-introduction seems to require the release of an adequate number of individuals. The precise number needed may be hard to assess, for it will depend on the species' patterns of fecundity and survivorship and on the impact of chance events. Despite this, it appears that a critical level of released animals must be exceeded for a re-introduced population to establish itself. As examples where this level was not reached, 15–20 Panamanian black-handed spider monkeys were released on Barro Colorado Island over 6 years in the late 1950s to early 1960s (Konstant & Mittermeier, 1982). In 1982 their numbers were still only 15–20. Again, in the pilot release of golden lion tamarins, 14 survivors of the quarantine period were released in mid 1984, of which 11 had either died or been removed one year later (Kleiman *et al.*, 1986).

Diverse experiences have led to recommendations about release strategies. In the effort to re-introduce the swallowtail butterfly to England, annual releases of 50–100 adults failed because those that survived one year were then swamped by the next release (Dempster & Hall, 1980). It was then proposed to re-introduce as many butterflies as possible in a single release to saturate the 30 ha habitat. A total of 228 adults was released, which laid an estimated 20000 eggs. Good caterpillar survival led to more than 2000 pupating but release even on this scale was inadequate.

The numbers of founders may partially explain the success or failure of a re-introduction. The numbers of sea otters seen in post-release surveys were significantly correlated with the numbers of animals released (Jameson *et al.*,

1982). In this instance, release of 20–50 otters annually at the same site over 3–5 years was recommended. From observation of dispersing beavers, a strategy was recommended whereby three to seven families or pairs should be released in an area of 20 km diameter, with a river system connecting all release and lodge sites (Zurowski, 1979). Other beaver re-introductions failed because starting numbers were too low to withstand the 40% losses through death and dispersal, despite breeding during the first year in the wild (Reichholf, 1976). Establishment of the European eagle owl depended on large initial releases and continuing annual releases totalling 550 owls over the first 15 years (von Frankenberg *et al.*, 1983). Some of this total represented deliberate overstocking to ensure the formation of compatible, breeding pairs. Repeated releases characterise successful re-establishment of birds of prey: over 9 years, 511 eastern peregrines were released to the wild (Cade & Hardaswick, 1985), and 82 sea eagles were imported to the island of Rhum, off Scotland, over 10 years (J. A. Love, personal communication 1987). Both have resulted in free-living populations which have started to breed. Where captive-bred animals are released, the requirement of repeated releases may guide breeding-programmes and determine the start of re-introduction. Release of the bearded vulture was delayed until a total of 10 captive breeding pairs, producing 10 young annually, was attained (Anderegg *et al.*, 1983), and the same principle governed the start of releases of both the eastern peregrine and whooping crane (Temple, 1983).

1.3.2   *Social organisation*

Many, if not most, translocations fail to establish new populations because of the stress on the animals, including the disruption of previous social bonds, which leads to lowered survivability or reproductive rate, or scattering in the new habitat and increased vulnerability to predators or other hazards (see, for example, O'Bryan & McCullough, 1985). Re-introductions, particularly of captive-bred animals, tend to be more carefully planned, with greater scope for objective selection of individuals on criteria of genetic background, sex and age composition. This selection is almost always the basis for release of animals in groups which mimic the natural social grouping. Experience has shown that this is probably the single most critical factor in attempting to ensure a successful re-introduction in any animals which are not asocial. For example, three wild troops of olive baboon, totalling 131 animals, were captured in Kenya and moved 190 km, to be released into areas already occupied by other conspecifics (Strum & Southwick, 1986). The translocations resulted in almost no mortality or post-release loss because movement of the entire societies was achieved after painstaking preparation.

Two examples demonstrate the consequences of releasing atypical social groups. Captive breeding of the pink pigeon resulted in a surplus of males, so that after release in Mauritius the population contained six males, two females and one of unknown sex (Todd, 1984). For a pair-forming breeder this re-introduction could not have been anything but a pilot effort. The golden lion tamarin has a complex social group, averaging 7.2 animals in wild groups, with a 1:1 sex ratio and no more than two adults of each sex. It is likely that only a single female breeds in a group and, at least in captivity, females exhibit very high levels of aggression towards each other, even if related (Kleiman *et al.*, 1986). The first tamarins to be released into the wild comprised a family group of eight and three male–female pairs. The performance and survival of the paired animals was so poor that future releases will only be of age-graded family groups, which will take advantage of the learning ability of the young and the breeding experience of the adults.

The inability of the spider monkeys on Barro Colorado Island to increase over 15 years led to the general observation that intact social groups should be re-introduced, or that artificial associations should be created in captivity which approximated the natural grouping (Konstant & Mittermeier, 1982). Thus, before empty otter habitat was restocked in England, a study on wild otters in Scotland showed that an adult male was likely to share its range with two breeding females (Green *et al.*, 1984). Accordingly, three unrelated cubs were removed from their mothers at 10 months, and kept together in an undisturbed enclosure until release at about 18 months old (Jefferies *et al.*, 1986). Intense monitoring showed that the group stayed in contact with one another during development of a home-range of 75 km², and one female had a cub within 12 months of release.

The development of suitable social groups in mammals in captivity is usually achieved merely through increasing familiarity between individuals, provided that the sex and age composition does not result in prolonged and damaging fighting in species with rigid hierarchies. If muskox are penned together with adequate feed, stable herd relations form which guarantee that the herd will stay together during long movements after release (Uspenski, 1984). Development of the stable herd may be so effective that when further muskox are released few are accepted into the established herd, thereby defeating the objective (Grauvogel, 1984). The Jamaican hutia is reared in captivity in pair-based, social family groups of up to 18 members, which were the basic unit for re-introduction to Jamaica (Oliver, 1985). The red wolf will be re-introduced into the USA in established mated pairs, which a pilot release study showed would stay together. In the past, the wolf was thought to travel in family groups, which the re-introduction may confirm (Parker, 1986). Wild-caught, adult beavers were paired in captivity in preparation for release (Zurowski, 1979).

The ability to mimic the natural social organisation in captive-bred founders for re-introduction depends on information acquired during planning stages. It emphasises the value of research studies on the species or related species in determining re-introduction techniques. In contrast to mammals, no attempt has been made to re-introduce a bird species with large, fixed membership groups, and because most birds are released into the wild as young, pair formation follows later without management assistance.

1.3.3   *Habitat modification*

Where the habitat requirements of a species are well known, the habitat may be improved to promote re-introduction success. This will be most feasible under two general conditions. The first occurs when exter-mination of the original population was due to habitat loss rather than direct persecution. Re-establishment of the large blue butterfly in southern England was assisted by increasing the grazing pressure from sheep and rabbits to restore the short turf. This encouraged the insect's food plant, thyme, and the nest density of the red ant species which was critical for larval development (Thomas, 1980). While such habitat modification may be most feasible for invertebrate re-introductions, the forest habitat of re-establishing golden lion tamarins in Brazil was protected from annual burning by firebreak construction and a large area of grassland limed to raise the soil pH to encourage forest regeneration (Kleiman *et al.*, 1986). The re-introduction of beavers into the Bavarian Danube was only feasible after their riverside habitat of poplars and willows had been restored through re-afforestation (Reichholf, 1976).

The second condition where habitat modification is practical occurs where otherwise suitable habitat is deficient in a specific structural feature. Thus, in contrast to the German example, beavers for release into Polish rivers were placed in artificial lodges (Zurowski, 1979). Similarly, the Jamaican hutia was released into areas transformed into suitable habitat by artificial warrens in the limestone. In other areas, natural fissures in the rock provided sanctuary to hutia colonies (Oliver, 1985).

Habitat management can also facilitate a beneficial behaviour. One of 16 rehabilitated chimpanzees marooned on an Ivory Coast island cracked palm nuts with a stone immediately on release. As the animals subsequently carried nuts from afar to the feeding station, more rocks were taken to, and dispersed around, the island, and most individuals adopted the habit (Hannah & McGrew, 1987).

Habitat modification in future may include the removal of exotic species, especially from islands, when they were the causes of decline in native species. Elimination of introduced Arctic foxes from several Aleutian islands preceded efforts to re-introduce the Aleutian Canada goose (Springer, Byrd & Woolington, 1978), and the elimination of exotic, predatory molluscs must anticipate re-introduction of captive-bred *Partula* snails to their Pacific island homes (Wells, 1988).

1.3.4 *Reproductive manipulation and genetic management*

Previous sections have shown the need in re-introductions, first, for releasing adequate numbers of animals and, second, for releasing animals of appropriate genetic composition. Many re-introductions or re-inforcements have used various techniques either to accelerate the production of animals for release or to influence their genotypes. Modern techniques in reproductive physiology, such as artificial insemination or embryo manipulation (see, for example, Samour, 1986; Hearn, 1986) are excluded here as they are not yet field techniques.

The present ability to modify breeding depends very much on the species or its reproductive system. Thus, reproduction in birds can be most easily manipulated because of their convenient habit of laying eggs (see, for example, Temple, 1978). Two intra-specific methods are common. The first is double-clutching, whereby first, or even second, clutches are removed, incubated and hatched (Dixon, 1986). If the nestlings are reared on a parental puppet, as in condors or cranes, the young birds are suitable for re-introduction (Bruning, 1983; Luthin *et al.*, 1986). The second method, fostering, enables greater interaction between an endangered wild population and captive-breeding programmes to re-inforce the former. Here, eggs or nestlings reared in captivity are inserted into suitable wild nests. Eagle owls reared more young in this way, provided the adult pair was given supplementary feed (von Frankenberg *et al.*, 1983).

Cross-fostering allows one species to raise the young of another. Thus, peregrine falcon chicks were placed in the nests of prairie falcons, to increase the former's numbers (Cade & Hardaswick, 1985). Sandhill cranes incubated and reared whooping cranes, and the foster parents taught the young birds new migration routes, but there is no evidence yet of whooping cranes reared in this way breeding successfully (Drewien & Bizeau, 1978). The value of cross-fostering in re-introductions is not yet clear (Temple, 1983). In contrast, cross-fostering in the wild may have saved one of the world's rarest birds, the Chatham Island black robin. By 1973, only seven individuals persisted in less than 5 ha of suitable habitat. In addition to translocating some to a safer island, the first clutches of the remaining two breeding pairs were removed, and second or third clutches were laid. Eggs were placed in the nests of the closely related Chatham Island tit, which incubated and reared the chicks to independence at seven to eight weeks old. Robins reared in this way subsequently mated successfully in their natural habitat (Flack, 1975). The equivalent technique for mammals is embryo transplantation, which is still far from a field method. However, it has been considered that red wolf embryos should be inserted into the congeneric coyote in the re-introduction area, with the presumed aim that wolves reared by coyotes would learn survival techniques more effectively (Parker, 1986).

Reproduction in re-establishing eagle owls was facilitated by releasing naive, single adult birds of appropriate sex into the territories of proven breeders who had lost their mates (von Frankenberg *et al.*, 1983). After capture and removal to productive lowland habitat, the flightless Lord Howe Island woodhen produced up to nine clutches in a year with supplementary feeding. In contrast, in its remnant high altitude habitat a bird usually produced one clutch of one or two eggs annually (Fullagar, 1985). Birds from this semi-captive-breeding were used to re-inforce the wild population most successfully.

While techniques for birds are more directed at increasing the numbers of individuals available for re-introduction, the emphasis with mammals has been on the genetics of potential re-introductions. Management of the golden lion tamarin in captivity was directed towards maintaining long-term viability through demographic and genetic management (Kleiman *et al.*, 1986). As the captive population increased fast while the remnant wild one decreased, it was apparent that genetic exchange between the two was needed through re-introduction, translocation and specific genetic intervention. This was the basis of a detailed breeding plan with the aim of producing the most suitable animals for re-introduction from a viable captive population. The world's Przewalski's horses have a breeding plan directed at reducing inbreeding levels through co-operative breeding agreements (Ryder & Wedermeyer, 1982). Horses from a secure, outbred population are the best prospects for successful re-introduction. Development of this type of comprehensive and accurate pedigree data for genetic management is more suited to mammals with small numbers of young than for birds.

While these techniques are still relatively crude, advances in reproductive technology will shortly offer the possibility for re-introduction of specified numbers of animals of prescribed genetic composition. This approach would then take advantage of the full array of genetic variation within captive populations.

1.3.5   *Supplementary feeding and release site fidelity*

In contrast to many translocations of wild-caught animals, re-introductions often involve small numbers of valuable, captive-bred individuals. Consequently, the main concerns on their release are, first, the animals' possible inability to find adequate food and water, and, second, that they may scatter or disperse far, resulting either in their death or an inability to locate and monitor them. Most re-introductions use supplementary feeding to provide food and to limit dispersal, employing a wide variety of techniques for various ends.

While some marooned populations such as chimpanzees rehabilitated on to islands require permanent provisioning to some extent (see, for example, Hannah & McGrew, 1989), for many species it self-terminates. Thus, golden lion tamarins had developed an adequate diet within 10 months of release (Dietz, 1985), and European otters caught enough fish after 12 days in the wild (Jefferies *et al.*, 1986). Chimpanzees on Rubondo Island in Tanzania found adequate natural food after 2 months because of the island's large size and particularly favourable forest resources (Borner, 1985).

Provisioning at the release site is an effective method of controlling dispersal through development of an area fidelity. This is the basis of the technique of 'hacking' birds of prey into the wild. In the case of captive-reared peregrines (Barclay & Cade, 1983), 4-week old fledglings are placed in a protective box at the release site and fed there until capable of flight at 40–45 days old. Food is supplied at the same site until the young birds have become self-sufficient and progressively disperse. Some peregrines raised on artificial towers become imprinted on them, resulting in their harassing younger birds subsequently being hacked from the same site. The presence of captive young sea eagles may lure in previously released birds (Nature Conservancy Council, 1985). Older birds may benefit newly released ones through fostering by providing wild-caught food. Provisioning at the release site and a captive decoy pair ensured that re-introduced pink pigeons returned and could be observed and even recaptured for blood sampling and re-banding (Todd, 1984).

Supplementary feed can also be used to promote dispersal. As the flight of young, captive-bred Andean condors became progressively stronger, their food was taken further from the release site until provisioning ceased when the birds were seen foraging and locating other feeding conspecifics (Bruning,

1983). The range of re-establishing griffon vultures in the Cévennes National Park of France was increased by dispersing feeding sites. The birds' dependence on provided offal decreased as they learnt to detect naturally occurring sheep carcasses, their traditional food (Fonds d'intervention pour les rapaces, 1985). Some released animals return to their release site even when their fidelity is not re-inforced with food. Three male cheetahs used an entire 1100 ha fenced reserve, but repeatedly returned to enter and mark their quarantine quarters (Pettifer, 1980).

Other techniques have been used to limit the dispersal of re-introduced animals. Indian rhinoceros were released into the 90 km² Dudhwa National Park, but an inner area of 27 km² was temporarily enclosed with electric fencing to prevent the animals from wandering into surrounding cultivation and to assist their establishment in optimal habitat (Sale & Singh, 1987). Confinement over three winters suppressed the natural desire of wild-bred European storks to migrate to Africa for the winter (Renaud & Renaud, 1984), provided extra feed maintained their condition through winter. This has resulted in a year-round population of storks in eastern France. Breeding pairs of eagle owls re-introduced into Bavaria were confined to an enclosure by clipping one wing of the female. The nestlings were fed by the male, and were free to disperse on fledging, leaving the pair to breed again at the same site (Holloway & Jungius, 1973).

Released muskox returned after their second winter on the tundra 50 km from the release site while it still had captive conspecifics (Uspenski, 1984). Shorter-term site attraction can be achieved by releasing only certain sex and age-groups of species with highly structured societies. After translocation of entire baboon troops to a new site, the subadult and adult males were confined to their cages for several days after the females and young had been released (Strum & Southwick, 1986). Combined with provisioning, this ensured that the females explored, and became familiar with only a limited area and returned each evening to sleep near the caged males.

A novel approach exploited the fact that the black rhinoceros defaecates in middens which are used to delimit territories. Hence, the droppings of each of a series of rhinos were trailed and left at points around a holding enclosure (Hillman, 1982). Despite no control experiment, each such animal spent the first few days of freedom in the close vicinity of its own dung. This technique might offer the possibility of encouraging limited dispersal by placing the dung of other rhinos near the release site, or for use with other species.

These examples show the prevalence of measures used in successful re-introductions to ensure that the released animals either remain near the release site or return to it at intervals. When dispersal is limited, the importance of locating the release site in good habitat is clear. Only if released animals are likely to explore further away is there a case for releasing them into suboptimal habitat. If the first founders then find optimal habitat, the release area will be vacant for subsequent releases without possibly adverse interactions between the two groups (1.3.6).

1.3.6   *Behavioural management and release techniques*

    When captive-bred animals are released into the wild, large behavioural changes will usually be necessary for survival. For example, diet selection in the wild presents a challenge as food items have to be sought, recognised or learnt to be edible; they must then be accessed and ingested. The extent of changes in foraging and feeding behaviour will depend on the species' trophic level and diet type, the diversity of potential food items and so on. In preparation for the wild, golden lion tamarins underwent thorough training to teach them to search for hidden, scattered foods rather than expect to find cut foods in traditional places (Kleiman *et al.*, 1986). After release the animals quickly developed a catholic diet, but survival time in the wild was strikingly correlated with feeding skill in captivity. Despite laboratory exposure, these tamarins showed poor evaluation of noxious or predatory animals, and the adults in particular were unable to plot a cognitive route through the forest which was structurally more complex than their zoo enclosure (Kleiman *et al.*, 1986).

    Pre-release training is most developed in the cases of the great apes such as chimpanzees and orang-utans. Successful rehabilitation for a free-living existence in the wild often depends on whether the animal was wild-born before enduring a period of captivity (Aveling & Mitchell, 1980; Hannah & McGrew, 1989). When single apes are to be rehabilitated, the effort required is prodigious and success at best marginal (Brewer, 1978; Mensink, 1986). Even if such individuals are rehabilitated successfully they must often be released into an area with wild conspecifics, with whom integration often fails (Aveling & Mitchell, 1980; Brewer, 1978).

    The same problems of integration and feeding techniques in the wild occur in large carnivores with group hunting methods. Teaching hunting competence to single lions can be prolonged, and their acceptance into wild prides uncertain (Adamson, 1986). In preparation for release, captive-bred red wolves, which hunt singly, will be increasingly given primary prey species, first as carcasses, and then as living specimens (Parker, 1986). However, captive-bred large carnivores with less social hunting techniques seem to adapt to hunting: captive-bred cheetah killed prey in a small enclosure at once, and the same three males were self-sufficient in the wild (Pettifer, 1980). Hunting behaviour may be partially innate in large carnivores, but early experience may be valuable: the first batch of nine wild dogs re-introduced into Hluhluwe Game Reserve, South Africa, comprised captive-bred and wild-born pups raised in captivity (Whateley & Brooks, 1985). The pack hunted in its pre-release enclosure and killed on its first day in the reserve (P. M. Brooks, personal communication 1986).

While partial habituation to humans in released animals may moderate dispersal after release, and certainly aids monitoring, overtameness may be disastrous. Cooper's hawks which had been frequently handled as nestlings suffered a significantly higher mortality after fledging than similar birds never handled (Snyder & Snyder, 1974). Experience with released birds shows that the avoidance of imprinting and of fearlessness towards humans is essential to improve survival in the wild (Temple, 1983). So, captive hatchling condors and various crane species are reared on puppets resembling at least a parent's head and/or in visual isolation from humans (Bruning, 1983). Over-habituation may represent a similar danger in the wild for mammals, but wild-type behaviour can develop very quickly: two adult female elephants, captured in Africa but then consigned to 14 years as trained animals in a USA zoo, were acclimatised and released into Pilanesburg Game Reserve in Bophuthatswana. Within weeks they behaved like wild elephants, avoiding human contact and protecting some 20 4–5 year-olds which they had adopted (J. L. Anderson, personal communication 1987).

Disease risk is a major factor in assessing the wisdom of population re-inforcements. Released animals must be free from infectious disease. Apart from jeopardising their own survival prospects, diseased animals may infect conspecifics or other species. This is a major risk where released animals re-inforce a wild population (Aveling & Mitchell, 1980). According to some opinion, re-inforcement of wild orang-utan populations should be discontinued because of the potential of introducing human diseases, amongst other concerns such as social integration and carrying capacity in the release area (Rijksen, 1978). At least one release of rehabilitated orang-utans has been cancelled because some had human tuberculosis (Jones, 1982). Stringent tests were made on pink pigeons before their release because of the occurrence of a highly endangered, remnant wild population 35 km from the release site (Todd, 1984).

The season at which animals are released can be critical for re-introduction success, with the main factors being seasonal range use, food availability or climate. On the basis of a limited experiment, red wolves will be released in spring, when more young and less wary prey specimens abound. This will allow the wolves 6 months' experience before conditions become harder

(Parker, 1986). Griffon vultures are released in the Cévennes when the bird-hunting season is over, and as near as possible to the start of breeding when weather conditions also reduce the risk of long flights from the release site (Parc National des Cévennes, 1981). With species such as mountain ungulates which have distinct seasonal ranges often at very different altitude ranges, winter release is preferred because winter habitat is more restricted in extent and locality (Geist, 1975), and any suboptimal habitat selection will affect survival more in the winter climate than in summer. Once such species as ibex and bighorn sheep have spent their first winter in suitable habitat, they will return to the same area the next winter after ranging away in summer.

In such instances, a traditional knowledge of seasonal ranges will develop readily. The problem is greater with birds that migrate much larger distances. In an effort to re-establish the Aleutian Canada goose on former island habitats, birds from the sole viable island population were added to the re-introducing birds on empty islands with the intention they should guide them to the California wintering grounds (Springer *et al.*, 1978). There was no evidence of any success. It has been proposed that Andean condors use the last vacated range of the Californian condor on a temporary basis until captive-bred numbers of the latter are adequate for release (Anon., 1986). Instead of removing the alien species before release of young California condors, the Andean could be useful instructors to speed adaptation to wild conditions and demands, if steps were taken to prevent any chance of interspecific breeding.

The techniques of actual release depend on the species being re-introduced, but all experience dictates that it should be as gradual and as stress-free as possible to promote successful transition to the wild (Elder, 1958; Mathews, 1973; Kear & Berger, 1980; Temple, 1983). Releases of pink pigeons were typical of birds, in that the cage door was opened with their usual feed placed just outside, and the birds still had free access to their cage (Todd, 1984). If animals are released in this way, with food available at their site of captivity in the release area, site fidelity is likely to develop. Translocated beavers were immobilised and placed in artificial lodges in the re-introduction area (Zurowski, 1979). As they recovered, the lodge was regarded as a haven, having the animals' own scent, and was then used to range from for days, months or even permanently.

Illustrative of many releases in translocation efforts, six adult and 30 yearling muskox were flown to a release site, and their crates arranged in a crescent with the doors facing inwards; when the doors were opened simultaneously the animals grouped and ran 3 km before halting. Any individuals separated during this flight could not be rounded up. Many muskox translocations have now shown that handling and release methods influence group cohesion on release (Grauvogel, 1984). In contrast, black rhinoceros translocations have shown that post-journey penning at the release site is much less stressful to the animal (Hillman, 1982). Compared to rhinos released without penning at the re-introduction site, such animals move away only half as far in the first week and month, and are more likely to establish ranges in the desired area.

A species' ranging pattern away from the release site and its fidelity to this site will determine the extent to which the same release facility can be used repeatedly. For the black rhinoceros it is recommended that small numbers only should be released from the same pen, and with long intervals in between (Hillman, 1982). This reduces the chances of aggressive interactions between residents and re-inforcing animals, or even between animals previously kept in adjacent pens. Release stations for red wolf pairs will be widely separated to avoid territorial conflicts early in the re-introduction (Parker, 1986). The location of second-year releases will be determined by the current pattern of territory location and use. This approach would avoid peregrine falcons returning to their release station and harassing younger birds being hacked there (Barclay & Cade, 1983).

1.4   Conclusions

Each case-history in this review of re-introduction efforts and techniques has lessons for re-introductions in general. A number of points stand out.

First, most genuine re-introductions use captive-bred stock. The individual history of each released animal, in terms of its genetics, rearing and experience before release, contributes to its survival prospects in the wild. This should influence captive-breeding policies and practices for any species that is ever likely to be a candidate for re-introduction (Frankham *et al.*, 1986).

Second, a high proportion of founders in many, if not most, re-introductions perish shortly after release. More surprisingly, there are no recorded examples of the opposite occurring, namely a population irruption of a species after returning to an otherwise complete community. This is in marked contrast to the irruptions to pest status of many introduced species. The performance of re-introduced species suggests that they are again soon subject to the aboriginal, environmental factors which regulated their numbers previously. This may be finely balanced, for the reindeer, *Rangifer tarandus* subsp. translocated to re-inforce the conspecific caribou, *R. t. arcticus*, of Alaska displayed an irruption typical of an introduction, while caribou numbers remained the same (Leader-Williams, 1988).

Third, re-introduction projects take time and, hence, considerable resources. Project duration depends on generation time, the number of founders, the criteria of success and so on, but the re-establishment of even a butterfly will spread over several years. Such time-spans are one reason why it is hard to say if some case-histories have actually been successful re-introductions.

Fourth, most successful re-establishments rely on comprehensive information on the species' biology, with insights into how the animal's behaviour can be manipulated or turned to advantage. This could be, for example, to obtain a successful release at the optimum season or a cohesive group in the wild. Furthermore, most successful efforts show evidence not merely of good scientific data on the species but of a deeper understanding of how naive, captive-bred animals will perceive and respond to their native, but novel, habitat. This understanding can be critical to success and has implications for project design and staffing.

As the guidelines for re-introduction procedures emphasise (1.2), a re-introduction project must consider biological factors without neglecting socio-economic, political, logistic and administrative aspects. In the story of the oryx re-introduction into Oman, these factors were either part of the feasibility, planning or execution stages or could be accommodated as they arose. In this sense, Oman's project learnt much from earlier re-introductions, but as the next chapters show, it also had some unique features.

2

The genus *Oryx*, and the Arabian oryx: biology, history and extermination from the wild

2.1   Introduction

This chapter has two main themes. First, comparative oryx biology is selectively reviewed, and then the fate of the aboriginal populations of the Arabian oryx is discussed. Their demise was the reason for the species' re-introduction into the deserts of Oman, without which this oryx would only exist in captivity.

As the techniques of re-introduction exploit the released species' ecology and behaviour, its biology must be understood. This chapter consists of a brief, and very partial, description of the Arabian oryx in its natural habitat. Such information underpinned design of the re-introduction methods (Chapter 5) and was the baseline against which the behaviour and performance of the re-establishing animals were assessed (Chapter 7).

This chapter also looks at comparative oryx biology for two reasons. First, the other oryx species provide clues as to Arabian oryx ecology or confirm scanty evidence, for its loss from the wild preceded any serious study of the species. Second, the Arabian oryx occupies more extreme habitat than the other species which, in turn, causes further adaptations and specialisation. Many of these were turned to advantage in the re-introduction, and its adaptation to extremes is pertinent to the general theory of re-introduction feasibility and methods (Chapter 10).

Analysis of the Arabian oryx's extermination from the wild shows the factors responsible, which must be known when assessing whether a re-introduction is feasible (1.2.1), and how it should be done. This chapter shows that many of the Arabian oryx's adaptations facilitated its extermination by humans, but these same factors were later exploited in management of the oryx for re-introduction.

2.2   The genus *Oryx*

Oryx belong to the Hippotragini, the family of horse-like antelopes. In addition to members of the genus *Oryx*, this group includes the roan, *Hippotragus equinus*, and sable, *H. niger*, antelopes which inhabit mixed woodland–grassland habitats in central, southern and partially into western Africa. The addax, *Addax nasomaculatus*, is also classified with the Hippotragini despite an anomalous appearance, largely because it shows many of the same adaptations as do oryx to very arid conditions in the Sahara desert.

Contemporary taxonomy allocates three species only to the genus *Oryx*, although there are five distinct forms, out of which only two have contiguous ranges. For present purposes these forms are referred to as species. Their historical distributions show that each is associated with an arid area (Fig. 2.1a, b): *Oryx gazella gazella*, the gemsbok, inhabits the Kalahari system; *O. g. beisa*, the beisa oryx, occupies the large arid area of Somalia and eastern Africa, with the closely related *O. g. callotis*, or fringe-eared oryx, in an adjacent range divided by the Tana river in Kenya; the scimitar-horned oryx, *O. dammah*, lives in the Sahara desert, while the Arabian peninsula is the home of *O. leucoryx*, the Arabian or white oryx. However, there are significant differences in rainfall (Table 2.1) and aridity (Fig. 2.1) between species' ranges. The gemsbok's range is mostly arid and semi-arid, with an annual rainfall between 50 mm in the driest part of the Kalahari-Namib desert to 250 mm in dry grasslands and bushland. In East Africa the fringe-eared and beisa oryxes inhabit arid and semi-arid zones, with high densities in bushland savannahs that have up to 300 mm rain a year. The two northern species had ranges into more arid areas: in historical times the scimitar-horned oryx ranged through the Sahara desert's central hyper-arid zone. Although principally an animal of the subdesert steppes of the arid and semi-arid zones, it migrated seasonally into the true desert and the southerly wooded steppes (Newby, 1974). In contrast, most of Arabia's interior, apart from its mountain ranges, is arid or hyper-arid because annual rainfalls average only 50 mm and many consecutive years may be rainless. The habitat of the Arabian oryx encompasses various desert types such as sand-dunes, stony plains and dry drainage courses known as wadis. Overall, this species lived in a consistently hotter and drier habitat than any other oryx.

The size and appearance of oryx are related to the aridity of their ranges (Plate 1 and Table 2.1). The various forms of *O. gazella* from less extreme areas are larger. The gemsbok is slightly heavier than the beisa and fringe-eared, and the latter are about 50% larger than the scimitar-horned oryx. This in turn is almost twice the size of the Arabian oryx. There is a further

28

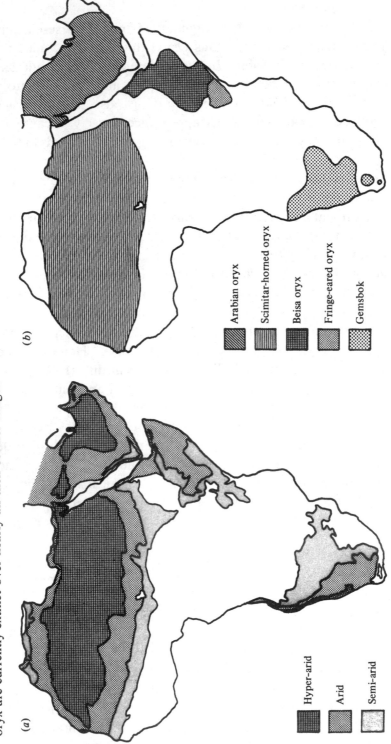

Fig. 2.1. (a) The distribution of arid zones in Africa and Arabia, from UNESCO (1979); (b) the historical distribution of the genus *Oryx* in Africa and Arabia. The Arabian oryx and scimitar-horned oryx are currently extinct over nearly all their former ranges.

(a) (b)

Plate 1. (a) Gemsbok in Botswana (photo: Anthony Bannister).
(b) Beisa oryx in Samburu Reserve, Kenya: note the abundance of trees
and ground vegetation. (c) Scimitar-horned oryx re-introduced into
Bou-Hedma Reserve, Tunisia (photo: Iain Gordon). (d) Arabian oryx,
in winter coat, in Oman walking across a typical stony surface of the
Jiddat-al-Harasis (January 1987).

(c) (d)

(a)

Plate 2. (a) An oryx herd basks in the sun on a cold winter afternoon (December 1986). (b) A herd shades under an *Acacia tortilis* on a tree-dotted plain. Even without a heat haze, the oryx in the shade are inconspicuous (November 1984).

(b)

Plate 3. (*a*) The herd male (far right) herds a newborn calf behind its mother, who walks and ruminates. The calf has just finished lying and has been joined by another calf a month older (far left), its nearest age peer in the herd. In the foreground is good winter grazing (December 1983). (*b*) A herd grazes on lush winter vegetation of annual and perennial grass clumps in a narrow *haylah* between stony ridges (foreground and background), with *Acacia tortilis* (November 1983).

(*a*)

(*b*)

(a)

(b)

Plate 4. (a) A Harasis campsite in a sandy *haylah* with shade from
Prosopis cineraria. The family has a frame dwelling and one of the
ubiquitous pick-up trucks. The goat flock is away grazing, leaving only
the kids nearby. (b) Harasis rangers on patrol watch an oryx herd next
to their vehicle.

cline in coloration and markings. The gemsbok presents a harlequin pattern of black and white patches against a sandy brown background. The besai does not have the continuous dorsal crest, nor black rump patch, and its tail is lighter in colour. The legs are less patterned, the flank stripe less accentuated, and there is less contrast between the flank and belly colours. The face markings are the same. The fringe-eared, not illustrated, has tufts on the ends of its ears, and has a basic brown body colour rather than the grey of the beisa. The scimitar-horned oryx has reduced face markings, with an off-white to white face, abdomen and legs, and a russet neck, upper chest and shoulders.

Table 2.1. *Comparative information on four oryx species and average annual rainfall in their habitats. Sources are shown in brackets, references below*

Oryx	Rainfall (mm)	Weight (kg)	Home-range (km²)	Group size	Density (oryx/km²)
Gemsbok	50–250 (b) (c)	m. 180 (a) f. 225 (b)	f. 400 (f) m. 20 (n)	36–50 (m)	0.13– 0.63 (n)
Beisa and fringe-eared	50–300 (b) (e)	m. 168– 209 (d) f. 116– 188 (d)	f. 200– 300 m. 150– 200^a (f)	av. 15–70 (f) range 1–120 (f)	0.07– 1.40 (f)
Scimitar- horned	< 50–200 (b) (b)	m. 135– 140 (g, i)	No data	30 (i) av. 15, range 6–28, temporarily < = 300 (l)	0.05 (l)
Arabian	< 50 (j)	m. 65–75 (j) f. 54–70 (a, h, j)	> 2000	1–15 temporarily larger (j, k)	0.05 (j)

^{a} Non-territorial males only; m, male; f, female.
Sources:
(a) London Zoo records, 1987
(b) Anon (1980)
(c) Mills & Retief (1984)
(d) Ledger (1968)
(e) van Wijngaarden (1985)
(f) Wacher (1986)
(g) J. M. Knowles
    (personal communication 1987)
(h) Thorp (1964)
(i) Newby (1985)
(j) This work
(k) Stewart (1963)
(l) Newby (1974)
(m) Dieckmann (1980)
(n) D. T. Williamson
    (personal communication 1988)

The Arabian oryx is a uniform white, while the flank stripe is absent or only an indistinct smudge. The lower limbs are a chocolate brown to black with the exception of pure white pasterns. There is merely a very light mane down the neck, and the black face markings are smaller in extent than in the gemsbok or beisa, although the genus' characteristic bar through the eye merges with a triangular patch of black below the ear and following the line of the jaw. Thus, the two oryx of drier areas show marked reductions in their patterning against lighter background coloration.

Despite these differences, both sexes in all species carry very long horns, which in the scimitar-horned oryx are recurved. Although females' horns are thinner, but sometimes longer than those of the males, who wear them down through thrashing trees as adults, the level of sexual dimorphism is very slight, and probably less than in any other antelope group. Males and females present almost identical silhouettes, but adult males weigh 10% more than females.

Herds in each species normally contain approximately equal numbers of adult males and females, with immatures (see, for example, Stewart, 1963; Dieckmann, 1980; Newby, 1985; Wacher, 1986). This may be an adaptation to nomadism (Estes, 1974). Bachelor herds, as in many antelopes, are absent or have been seen rarely (Walther, 1978). Under good habitat conditions, single territorial males are seen in the gemsbok (Estes, 1974; Dieckmann, 1980) and fringe-eared oryx (Wacher, 1986). Single males and the dominant herd male defaecate in a squatting position, depositing their faecal pellets in a pyramid rather than as a scatter over the ground. These pyramids are most common at intervals down tracks or in conspicuous open patches near favoured shade trees.

A dominance hierarchy involving both sexes is found in herds of each species (Hamilton, Buskirk & Buskirk, 1977; Stanley Price, 1978 and personal observations; Wacher, 1986). The lack of sexual dimorphism is probably due to the female hierarchy and the intergroup intolerance observed. The striking markings and colour patterns of the East African oryx may be a means of promoting individual recognition and conveying efficiently the content of interactions between animals (Kingdon, 1982). Consistent with this, a wild herd of the more colourful and patterned gemsbok presents a picture of ceaseless low- to medium-level aggressive interactions, at higher rates than in the fringe-eared oryx (Stanley Price, personal observations). The virtually monochrome Arabian oryx naturally feeds more spaced out, but often only one or two low-level interactions are seen in an hour between members of an established herd. Its face markings are far less variable than in the fringe-eared, where they can be used to recognise individuals. Thus, while the

Arabian oryx has a dominance hierarchy, its coloration has evolved for other, or additional, functions.

The oryx is renowned for its ability to survive in hot, dry areas without drinking water for months, and the recorded patterns of distribution in the Sahara (Newby, 1985), the Kalahari (van der Walt *et al.*, 1984) East Africa (Kingdon, 1982) and in Arabia (Stewart, 1963) all confirm that drinking must be rare. However, even wild-caught Arabian oryx drink water when offered (Stewart, 1963). Experimental work under natural and semi-natural conditions with the fringe-eared oryx show that it has a low water requirement (Stanley Price, 1985*a*, *b*), which can be met by large drinks at infrequent intervals, and it has a lower rate of water turnover than the camel (King, 1979).

Research on the fringe-eared oryx has shown some of the physiological bases to its tolerance of hot, dry environments. When mildly dehydrated, oryx body temperature rises to 45 °C in the face of high air temperatures, and this accumulated heat is dissipated passively at night (Taylor, 1969). Damage to the heat-sensitive brain tissues is prevented by a counter-cooling system at the base of the skull (Taylor & Lyman, 1972). Evaporative water loss by oryx is very low, and when dehydrated above 41 °C the oryx pants, cooling itself for the loss of 60 % less water than would a sweating, normally hydrated animal. These facts suggest that a variable body temperature is a fundamental oryx strategy, and that a state of mild dehydration may bring into operation a suite of water-conserving mechanisms.

2.3   Oryx movements and home-range sizes

Observations on all oryx species agree that they are mixed feeders, with the bulk of the diet as relatively coarse grass, augmented by forbs, mostly ephemerals (Stewart, 1963; Field, 1975; Mills & Retief, 1984; Newby, 1985; Stanley Price, personal observations). Shrubs and tree browse are rarely taken, although the Arabian oryx takes newly flushed *Acacia* spp. foliage in the absence of green grazing (T. H. Tear, personal communication 1987).

The grazing resources of oryx habitat depend on rainfall, and for the four oryx species there is a general relationship between increasing aridity and increase in home-range size (Table 2.1). There are no precise data for the scimitar-horned oryx but as grazing areas may be 300 km apart in Chad (Newby, 1985), ranges of several thousand square kilometres are likely, as in the Arabian oryx. Group sizes in the two larger species of the less arid areas are also larger. Oryx densities show the same parallels (Table 2.1). The true desert species live at low overall densities: the scimitar-horned in Chad occurred at around 0.05 oryx/km² (Newby, 1974), and the re-introduced

Arabian oryx in Oman also lives at 0.05 oryx/km² (6.3.3). In contrast, the Kalahari gemsbok lives at up to 0.63 oryx/km² (D. T. Williamson, personal communication 1988), while maximum densities of 1.4 oryx/km² occur in the best East African habitats (Wacher, 1986).

All oryx species move over great distances in response to rain, and the largest groups occur as temporary aggregations on localised areas of fresh grazing (Stewart, 1964; Gillet, 1965; Newby, 1974; Mills & Retief, 1984; Wacher, 1986). The scimitar-horned of the south side of the Sahara desert made regular movements through the Sahel following rainfall caused by passage of the Inter-tropical Convergence Zone (Newby, 1974), although local distribution depended also on the availability of shade and the presence of humans. Until the mid 1970s hundreds of oryx aggregated on these migrations. In southern Arabia, in contrast, rainfall is less predictable without indication of any gradient of increasing precipitation south of the Empty Quarter margin. Here, the oryx followed rainfall adventitiously, preferring the stony plains or *jol* at the southern fringe of the sand-sea, although they penetrated the dunes when their grazing was good and shading (i.e. resting in the shade of the trees) was not necessary (Stewart, 1963). Summers were spent, perforce, on the *jol* because of its rock shade. There are no records of Arabian oryx occurring in numbers or densities or with movements comparable to those of the scimitar-horned. Its habitat is, effectively, more arid than those of the other species, and its smaller body and group size represent some of its adaptations to this extreme environment.

2.4   Adaptations of the Arabian oryx to the desert

The last section showed that oryx are generically adapted to arid areas. The Arabian oryx shows further specialisations, which are described in this section because they are directly relevant to the re-introduction rationale or methods. Unless attributed otherwise, all observations were made on the oryx re-introduced to Oman.

2.4.1   *Thermo-regulation*

Seasonal temperature oscillations in Oman (4.4) require oryx to avoid absorbing heat or to dissipate it in summer, while winter temperatures drop below the probable lower limit of their thermo-neutral zone. In the four hottest months of the year, day temperatures exceed a basal body temperature of 39 °C, while at night they are lower. In this period the gradients of heat flow between oryx and the atmosphere reverse twice daily.

Seasonal coast changes reflect these demands. The adult summer coat is short, sparse and highly reflective. All markings are brown rather than true black and the leg patches show less contrast against the white. In some animals the flank stripe is absent in summer. When both maximum and minimum temperatures decline rapidly in October (Fig. 4.4), the fur grows longer and thicker, and may take on a yellowish tinge. The areas of black on the leg spread and all markings darken (Plate 1*d*).

Variations in behaviour and activity patterns enhance the effects of these coat changes on heat balance. In summer, the oryx may shade from 08:00 to 18:00 h, depending on circumstances, but they must then feed at night. In winter, the animals are active throughout the day, lying from sunset to dawn in the lee of a tussock of grass or sand mound to escape cold winds. On cold mornings oryx stand motionless, oriented into sun as they pilo-erect, giving a suede appearance and allowing solar radiation to strike the black skin under the white fur. On cool afternoons oryx bask in the open (Plate 2*a*), appearing to absorb heat before the long, cold night, for which the more extensive and intensive winter black areas are probably an adaptation. As a white animal reflects more incoming radiation, while a black one absorbs more heat (Finch & Western, 1977), the seasonal changes in coat and coloration have a rational basis. Seasonal shifts in activity are augmented by swift responses to short-term climatic change. For example, if a cloud shadow passes over a shading herd, all may start grazing in the open almost immediately. Thus, oryx skin, coat texture and colour, and the animals' behaviour, are all finely tuned to maintaining heat balance. Achieving this balance may be the most efficient means of water conservation without recourse to more extreme physiological strategies and highly modified behaviour.

2.4.2   *Conspicuousness and crypsis*

Large antelopes without striking sexual dimorphism may exploit their silhouette and patterned markings for conspicuousness (Estes, 1974). The Arabian oryx is the whitest and least patterned oryx, and with the sun approximately behind an observer a single animal in the open is clearly visible to the unaided human eye up to 3 km distant. Many reports of the species comment on its visibility from a distance because of the highly reflective pelage (Thesiger, 1959; Stewart, 1963). Carruthers (1935) found the oryx 'glistening white in certain lights... but at other times with a change of light they were extraordinarily difficult to see'. Against the sun, the oryx can be invisible at 100 m because of glare and lack of reflection from the animal, but this occurs over fewer combinations of angle to sun and elevation than those rendering the oryx highly visible. Oryx take advantage of their con-spicuousness when searching for a herd by standing motionless on prominent ridges to advertise themselves (6.6.1).

The white colour has, perhaps, an even more important function. When an oryx moves into shade, particularly if dappled, it again becomes very hard to see (plate 2b). This photograph was taken mid-afternoon on a clear and warm winter day. In summer, shading oryx are virtually invisible because of a combination of effects: a ubiquitous 'white' light due to atmospheric dust and heat shimmer reduces visibility and definition, and the oryx often lie in shallow scrapes under trees with their horn tips lost against the canopy. A darker animal would be less cryptic in the shade. The advantage to the oryx of being cryptic is that in summer most daylight hours are spent shading, even under moderate conditions (6.2.1). Despite very acute eye-sight allowing them to detect potential predators from afar in the open, shading oryx are most reluctant to leave shade, even in a hunted population when the source of danger must have been seen for some time (Stewart, 1963). This reluctance to move is probably a response to the potentially fatal problems of thermo-regulation if an oryx has to run when it should be in the shade. After a few years in the desert, re-introduced oryx on their own exploited their crypsis in the shade: on several occasions, after a single oryx had been sighted from a distance by a vehicle patrol for the first time for weeks or months, the patrol then found the area apparently empty. Only very close searching revealed the oryx had deliberately moved under an *Acacia tortilis* on hearing or seeing the vehicle (T. H. Tear, personal communication 1988).

Arabian oryx calves are a sandy-brown colour for the first 2–3 months of their lives, in common with the other species. This crypsis is an obvious adaptation to reducing predation risk during their lying-out phase (2.4.4). The more desert-adapted Arabian and scimitar-horned oryx have evolved their conspicuousness as adults for specific purposes and because predator pressure in their habitats is much reduced through fewer predator species and lower densities of prey. While the patterning of the gemsbok and beisa oryx serve short-distance communication between animals, the Arabian oryx has become white for thermo-regulation and as an aid to locating one another over larger distances and at lower densities.

2.4.3   *Locomotion and endurance*

Although fringe-eared oryx have large hooves (Kingdon, 1982), these are even more exaggerated in the desert species, especially in the Arabian. Adapted to soft, sandy or loose, stony surfaces, the animal leaves a track in which the entire outline of the hoof is imprinted to a uniform depth. This oryx is most suited to a flat terrain, where its normal walk is 'cow-like' (Carruthers, 1935) (Plate 1d) and less graceful than that of the beisa (Stewart, 1963). Its flight when chased is slow and cumbersome, with no sharp turns as in the beisa (Stewart, 1963), and in Oman the species shows a most marked reluctance to move at anything but a walk. Running plays no part in normal social life and interactions, and the oryx cannot change quickly from walking to running.

Running oryx display little stamina (Kingdon, 1982) and in open country the Arabian oryx can be chased by vehicle and caught surprisingly quickly (Grimwood, 1967). On the other hand, it is adapted for walking great distances. A male was recorded wandering at least 93 km in 24 h (Grimwood, 1962), while a pair in Oman travelled 70 km between 16:00 h and 08:00 h the next morning. This endurance when walking enables oryx to move between widely spaced pastures, although they may also break into a ponderous canter for stretches of 0.5 km when on long journeys. Low predation pressure in the desert may have allowed adaptations for walking stamina to develop without penalising predator-escape behaviour. The Arabian oryx ruminates when on the move across barren areas, rather than delaying this activity to rest periods as do most ruminants. Periods of rumination and walking alternate rapidly with taking a few mouthfuls from an isolated grass clump.

2.4.4 *Behaviour and temperament*

Female oryx of all species have a *post-partum* oestrus, so that from the birth of her calf a female is consorted by at least one adult male, whether the birth occurred in a herd or in isolation. The consort relationship coincides with the calf's lying-out phase during its first month, when most of the day is spent lying, largely hidden and motionless, and its periods of activity are short. The Arabian oryx shows a unique behaviour associated with the consortship. It was witnessed after each of 30 births in Oman and, in each case, the consort was the herd's dominant male. As soon as a lying calf rises to approach its mother, the male immediately takes up a position a few metres behind the calf and, with his head low to the ground, he growls and directs it straight to its mother (Plate 3a). When suckling is complete, the trio moves off, invariably with the female leading, alert and looking forwards, with no backward glance to where the male is usually closer to the calf and again herding it after her. Often, the male apparently 'decides' when the calf should lie again, for he gently knocks it flat with the base of his horns. When the mother turns and attempts to return to the sprawled calf, he drives her away. Both may then stand 30 m away, watching while the calf rises and wanders off to a suitable lying place. Thus, both adults know the calf's lying out site, and stare at intervals in its direction as they move away 1 km or more over the hours that the calf lies.

In almost every case, the attending male was the dominant herd male and, hence, the calf's father, but a newborn calf elicits the same behaviour even if sired by another male. This occurred after a change of herd male during the female's gestation. The consistent appearance of this behaviour suggests that it promotes the calf's survival, and its unique occurrence in the Arabian oryx (although information on the scimitar-horned is too scanty) suggests a further desert adaptation. The chances of separation of mother from calf are probably not great in open desert and are not reduced by the presence of the male. On the other hand, avoidance of calf predation might depend on detecting the predator as far away as possible. The mother is able to devote her attention to this, knowing the male is guiding the calf after her. While a young fringe-eared calf hopes to avoid detection by a predator by lying as the adults flee (Kingdon, 1982), there is less chance of lying undetected amongst the sparse desert vegetation once spotted. This behaviour by the male may therefore have evolved to improve the female's early-warning capability. Protective behaviour by dominant male sable antelope is recorded (Estes, 1974), and there is evidence that these males are likely to be the calves' fathers (Ross, 1985), but such solicitous behaviour does not develop invariably nor occur so consistently over a calf's first month as in the related Arabian oryx.

Although Arabian oryx are recorded as impaling on their horns predators, such as the caracal (Crouch, 1962, in Stewart, 1963) or Arabian wolf (Harasis observation), the species is less aggressive than the fringe-eared, and newly caught animals quickly resign themselves to their condition (Stewart, 1963; Nokes, 1965) and tame easily. Two females walked, lightly roped together, at the head of a flock of sheep 500 km from Riyadh to Kuwait (Hamilton, 1918).

Fighting between male Arabian oryx occurs under natural or semi-natural conditions only on the arrival of a strange male in a herd. The fights last only a few minutes, and repeats are rare. Prolonged and damaging fights are common in the other species, and this lack of fighting by the Arabian is probably responsible for the less massive neck and horn roots (Plate 1), which distinguish its profile. The Arabian oryx neck drops in front of distinct withers, unlike the gemsbok and beisa oryx, which have very broad tendinous attachments between the thoracic vertebrae and the rear of the skull.

2.5   The oryx in Arabia, its decline and extermination

Oryx sightings have been collated to obtain a picture of the species' progressive range reduction in historical times (Stewart, 1963). The original distribution is assumed to have covered the whole of the interior of peninsular Arabia, except where the terrain was obviously unsuitable, and extended into

Syria. The species' historical distribution and its erosion in this century is shown in Fig. 2.2. By 1900 the population range had severely contracted, and by 1917 the oryx occurred in a northern population in the Great Nafud and a southern one in and around the Rubʾ al Khali of southern Arabia. The northern population became extinct around 1950.

The once-abundant oryx along the northern edges of the sand-sea in south-west Saudi Arabia and the modern United Arab Emirates were eliminated in the 1930s. The population between the western end of the Rubʾ al Khali and the North Yemen mountains followed in the 1950s. There then remained oryx only through the long southern margin of the sands, from the Aden Protectorate (now the People's Democratic Republic of Yemen, PDRY) into Oman. The area of this population was progressively reduced from both ends: no oryx survived west of 47° E by 1951, or west of 51° E by 1962. By 1970 a sighting of four oryx may have been the last as far as 52° E (R. H. Daly, personal communication 1987). At the north-eastern end of the range, oryx

Fig. 2.2. Distribution and progressive decline of Arabian oryx this century. Dotted lines enclose Arabia's main sand areas. Redrawn from Stewart (1963).

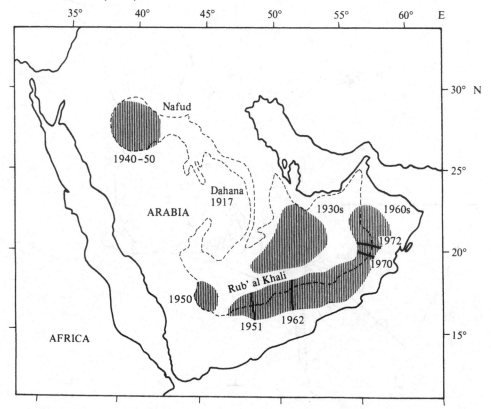

survived in Oman's interior until the mid 1960s. After that date, the entire world population was confined to an area of central Oman.

This sole inhabited oryx range was an area called the Jiddat-al-Harasis, which became the re-introduction area of the 1980s. Through the 1960s, various accurate reports of sightings were made (Fig. 2.3). Several herds of 20–25 beasts were seen in 1958–60 in the area defined in Fig. 2.3 (Woolley,

Fig. 2.3. Sightings of oryx on the Jiddat-al-Harasis in the late 1950s and 1960s. (See table opposite.)

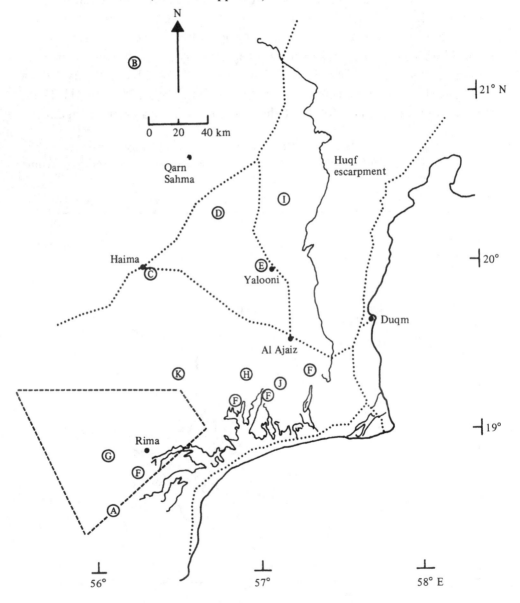

1962). From this, combined with recollections of local Harasis tribesmen of the areas favoured by oryx or where animals were seen, it is clear that the population in the first half of the 1960s was substantial. After travelling in the Eastern Aden Protectorate and enlisting the assistance of many people familiar with the Jiddat-al-Harasis, Stewart (1964) considered the latter might have 100–200 oryx. Traverses on camel-back through the area suggested that 100–400 oryx survived (Loyd, 1965).

Assessments of the threats to the remnant population differed. The remoteness of its location might have protected the population for many years (Woolley, 1962) but the persistent and skilful hunting of the *bedu* tribesmen was seen as a clear threat (Loyd, 1965). To others the future of the population was precarious in the face of highly organised, motorised hunting parties (Stewart, 1964). M. P. Butler (unpublished diary 1985, and personal communication 1987) travelled with camels across the Jiddat-al-Harasis in late 1965, saw oryx and warned the International Union for the Conservation of Nature and Natural Resources in 1966 and the Fauna Preservation Society in 1970 of their impending extermination by foreign parties. The causes of this last population's demise are discussed below, but in 1972 the last herd of six oryx was eliminated on about 18 October (Henderson, 1974). Since then,

Sightings of oryx on the Jiddat-al-Harasis in the 1960s. Sighting letter refers to map opposite

Sighting	Year	Oryx	Source
A	1958–60	Several herds of 20–25 One caught here 1960	Woolley (1962)
B	1963	2 adults, 1 calf shot	Loyd (1965)
C	1963	1 pregnant female shot	Mustoe (1964)
D	1963	12, including 4 young	Mustoe (1964)
E	1964	2 adults, 1 calf	Loyd (1965)
F	1964	4 groups of 4, 3, 2, 2 oryx seen by Longdon	Loyd (1965)
G	1965	1 male shot	M. P. Butler, unpublished diary (1965)
H	1965	5 adults, 1 calf	M. P. Butler, unpublished diary (1965)
I	1965	4 adults, 1 calf 2 adults, 1 calf	Nokes (1965)
J	1965	6, including 2 calves	Nokes (1965)
K	1972	Last herd: 6 killed or captured	Henderson (1974)

there has been no confirmed sighting of an aboriginal wild oryx in all Arabia.

2.6   The causes of extermination

The causes of the oryx decline and eventual extinction had a direct bearing on the feasibility of the re-introduction. Four main types of offtake were recorded.

(1) Desert tribesmen have always hunted the oryx, even if their earliest methods remain unknown. In the recent past, success in open country required great skill and ingenuity, with only an ancient rifle and a camel to hide behind. Many men might never kill an oryx in their lives, but those who had done so gained status in their society (Talbot, 1960). Thomas (1932) describes some of the uses of the oryx and its supposed virtues, while older Harasis (personal communications 1980–6) confirm that the flesh was the most highly prized meat, with the skin, horns, internal organs, rumen fluids and solids all used.

Foreign explorers and travellers also shot oryx for meat (see, for example, Thesiger, 1959; Thomas, 1932), but the hunting efficiency of their parties was probably the same as that of local tribesmen.

(2) Areas of oryx habitat became accessible and known to outsiders. In Arabia, the impetus for much road and water-well development has been oil exploration, and, even if this phase was unsuccessful, its infrastructure remained in place. In an area as featureless to an outsider as the Jiddat-al-Harasis the existence of one graded road across it largely removed the need for local guides. Oil industry parties were responsible for killing oryx on occasion (see, for example, Anon., 1979), but records are understandably few. On the Jiddat-al-Harasis, some tribesmen employed by exploration parties were tempted, under the influence of others, into chasing oryx in company trucks.

(3) The most commonly invoked cause of oryx decline was motorised hunting parties of Arab origin. Such expeditions comprised large convoys, with fuel and water bowsers, and the objects were the hawking of houbara bustards and hares, and the chasing of any gazelles or oryx encountered. The vehicles and support facilities allowed a very large area to be covered every day, with efficient removal of its wildlife.

In 1961 the eastern portion of the Aden Protectorate was raided by a hunting party which had crossed much of the sand-sea of the Rub᾿ al Khali, and it then killed 48 oryx, approximately half of the supposed population (Grimwood, 1967). Further west, the population of central Oman was also harried by such parties. The Harasis, who had not one vehicle between them, were powerless to do anything, and, in any case, the financial lures for providing guiding services in an area of meagre employment opportunities were overwhelming.

(4) Efforts to capture live oryx resulted in the removal of some and the death of others, by people with various motives. For example, a Saudi Arabian party came across a herd of 36 oryx in the Rub᾿ al Khali in 1951 and captured two successfully (Anon., 1952). In 1961 two of the remnant Aden population were captured but both died (Grimwood, 1962). Operation Oryx (3.2), caught four oryx from the same area in the following year.

Shaikh Kasim bin Hamid from Qatar visited an area between the Empty Quarter and the Oman mountains each winter, with the stated aim of catching fresh stock for his herd (R. Izzard, personal communication to the Fauna Preservation Society 1969), but found the population, which he estimated had been 200, was extinct in 1968. He blamed the followers of new oil camps in the area, from whom he had always kept oryx locations secret. The Harasis remember from the early 1960s large parties from Qatar chasing and catching oryx in nets on the open west Jiddat-al-Harasis (4.2). Large groups were loaded at a time on to trucks for the 900 km overland journey back to Qatar.

An oryx captured on the Jiddat-al-Harasis in 1960 subsequently became a founder of the captive-breeding endeavour (Woolley, 1962). Another expedition to the same area caught a 9-month-old female calf to be a regimental mascot by permission of Sultan Said bin Taimur (Nokes, 1965). This expedition was hailed at the time as a great conservation achievement because of its first photographs of wild oryx. However, the photographs show a running oryx, exhausted by pursuit, and the account omits mention of the first captured oryx dying from exhaustion and a second shot from the same herd for meat (Harasis, personal communications 1987). From fresh tracks, Henderson (1974) describes how, of the last herd of six oryx, three were captured alive and three were killed, one abandoned after being run to exhaustion and struck by a vehicle, while one adult and calf were eaten. It seemed that the captured oryx were either to replace animals in the captive herds of the Arabian Gulf, or for trading as the species was suddenly valuable through its scarcity in the wild.

No doubt, many motorized parties set out to chase oryx both to kill and capture. The conservation aims of those who wanted to catch oryx may have been sincere, but their techniques of capture and care were often inadequate, although the Arabian oryx is highly stress-tolerant and easy to manage (Grimwood, 1967). In addition to the direct removal or death of animals, many more may have died later through heat exhaustion or an irretrievably upset water balance through being chased. Thus, all motorised parties must have had a severe impact directly and indirectly on any oryx herds found, and thence on the population.

Even though Loyd (1965) was told that Harasis hunters had killed five oryx in three months of 1964, it is unlikely that their hunting pressure would ever have eliminated the population. The evidence that motorised parties were responsible seems conclusive. This is supported by the progressive loss, from around 1945, of the northern oryx, then those from the north side of the Rubʾ al Khali, and finally from east and west of the sands' southern margins, until the oryx remained only in the most remote and least accessible part of Arabia. Even this vast refuge was inadequate in the face of the pressures of the 1960s and early 1970s, despite a local tribe who wished to preserve the oryx.

The final demise of the Arabian oryx in the wild is now being closely paralleled by the scimitar-horned oryx. Despite a vast historical range across the Sahara and Sahel (Fig. 2.1), its population is now so reduced that it occurs in small numbers in a few areas only. The principal causes of its decline are the same as for the Arabian oryx, with the added ingredient of civil unrest and war (Newby, 1985). It is ironic that, as Africa's deserts spread, the scimitar-horned oryx, so adapted to these habitats, is highly endangered, while the less arid-adapted beisa oryx and gemsbok have large populations over extensive ranges.

2.7   Conclusions

This chapter provides no more than a brief and selective survey of oryx biology to show that the genus *Oryx* is adapted to arid lands and that the Arabian oryx has evolved a variety of characteristics to meet the demands of its more extreme environment. The Arabian oryx is a desert specialist amongst oryx, in its heat tolerance and water independence in the hottest season (6.9). Its relative lack of aggression and the consequent physical changes, the low rate of interaction and its reluctance to move at anything other than a walk suggest that it has become an energy-conserving species. In the wild it has not been seen to exhibit the ritualised, tournament running display, in which entire herds of fringe-eared oryx indulge under certain conditions (Kingdon, 1982). Occupying areas which may receive no rain for 5 years (4.4), food supply may have limited population size naturally, as with other vertebrates in such environments (Wagner, 1981). The white coat colour confirms that natural predation has been a minor factor in Arabian oryx evolution, although making it very vulnerable to predation by modern man in vehicles.

As many of these desert adaptations are suggested from mere observation of re-introduced, zoo-bred animals or their offspring, the species must possess many physiological adaptations, similar to, more developed or in addition to those seen in other oryx species, but all awaiting discovery. These observations partially anticipate the evidence of later chapters that captive-bred, re-introduced oryx are capable of responding to stimuli such as distant rainfall, and can learn to exploit their ancestral environment for their survival. The lack of such ability might have been fatal to re-introduction success.

3

Operation Oryx, captive breeding and re-introduction planning

3.1   Introduction

The Arabian oryx was totally exterminated from the wild in 1972. No re-introduction would have been possible without the existence of captive, breeding populations before this date. The capture of wild oryx to establish some Middle East populations undoubtedly contributed to this extermination and, moreover, the performance of these herds was often indifferent. The most effective measures to save the species were being taken in the Western world. These followed the mounting of Operation Oryx, whose aim was the deliberate removal of some of the last surviving wild oryx to establish a viable captive herd, with the longer-term objective of re-introducing the species to its original habitat in Arabia.

In this chapter, the results and consequences of Operation Oryx are described as part of the sequence of wild animal capture → captive breeding → re-introduction to the wild. The earlier phases are directly relevant to the Oman re-introduction because the latter depended on captive-bred animals. Moreover, performance of these oryx in the wild might reflect their management, especially genetic, and the number of generations spent in captivity. Such aspects, and the overall relationship between the parties involved in the different phases of the conservation sequence are relevant to any other re-introduction scheme.

3.2.   Operation Oryx

The first comprehensive account of the problems facing the Arabian oryx in its native range predicted that the decline in numbers would progress inevitably to extinction within a few years (Talbot, 1960). Capture was deemed a more urgent priority than an ecological study of the wild oryx.

Some oryx should be removed for captive breeding, which would not be difficult with catching-vehicles and air support for spotting.

A little earlier, in 1959, the Fauna Preservation Society had been informed by M. A. Crouch of a population of 80–100 oryx in the area of the Wadi Mitan in the Eastern Aden Protectorate, where it borders with Oman. Its viability was suddenly jeopardised by reports of a hunting party which reached the area in December 1960 from across the Rub? al Khali, and killed at least 48 oryx. The Fauna Preservation Society then took the initiative to mount an expedition to catch enough of the surviving oryx to start a breeding herd, with re-introduction as an eventual aim. The expedition was raised by Maj. I. R. Grimwood, and it took place in April and May 1962, despite near-cancellation due to a report in February 1962 that the raiders had returned to the Wadi Mitan and killed another 16 oryx.

The preparations and activities of the Operation Oryx team have been described in detail (Grimwood, 1962; Shepherd, 1965). Despite many problems, the expedition succeeded in catching four oryx, three males and a female in the vicinity of Wadi Mitan, right on the Eastern Aden Protectorate–Oman border. The animals were intercepted as they made their southward, spring move across the stony plain from the sand seas to spend the summer in the more shaded *jol* (2.3). The second captive, a male, died from capture stress after a long chase through difficult country, but was found to have a recent bullet wound in the leg. After an exhaustive search of an area of 15 000 km² the expedition concluded that a total of 11 oryx had survived the hunting parties. It had captured four of these and seen a further two which could not be pursued. Five others had moved east further into Oman where pursuit was not sanctioned, and two had been alarmed by the vehicles or aircraft and had slipped into the sand-seas. Although this population had clearly reached the brink of extinction, it persisted for a few more years; four oryx were seen in 1970 at 51° 58′ E, 18° 30′ N in the Wadi Mitan of Oman, only 50 km from the area of Operation Oryx eight years previously (R. H. Daly, personal communication 1987).

3.3   Establishment of the World Herd of Arabian oryx

The three oryx from Aden were flow immediately to Kenya, where they spent a few months in specially prepared quarters in the semi-desert north. However, this was not a long-term home because of the risk of foot-and-mouth disease in the area.

These animals were the property of the Fauna Preservation Society (FPS), to which the Zoological Society of London (ZSL) agreed to add on breeding loan the single female it possessed. This animal had been caught in the Jiddat-al-Harasis of Oman in 1958 (Woolley, 1962), and arrived in London in 1962. In late 1962 the FPS proposed that the four animals should be split as two pairs to be bred at London and Basle zoos. The members of Operation Oryx protested vigorously at this evident violation of the spirit and aim of the expedition, namely to set up a breeding herd in an area of climate approximately similar to that of Arabia. In the meantime, Grimwood had been offered a home for the oryx in the Maytag Zoo at Phoenix, Arizona, where he described the climate as ideal and the facilities impressive. Moreover, the Shikar-Safari Club in the form of its president, the late M. A. Machris,

Table 3.1. *Main aspects of the World Herd ownership agreement in 1964, from the Fauna and Flora Preservation Society*

An agreement on the ownership of each individual animal in the Arabian oryx herd established by the Fauna Preservation Society in co-operation with the Arizona Zoological Society, the Shikar-Safari Club, the Zoological Society of London and the World Wildlife Fund.

Principles
(1) Every animal must be named and tagged and its ownership decided upon immediately it joins the herd.
(2) Every newborn animal must be likewise named, tagged, and its ownership decided upon within 24 hours of birth.

Rules: to fix the ownership of animals born into the herd
1. The first animal born alive to any female and which lives for 24 hours becomes the property of the owner of its mother.
2. The second animal born to any female becomes the property of the owner of its father.
3. The third animal born to any female becomes the property of the breeding establishment.
4. The ownership of the fourth, fifth, sixth and subsequent animals born to any female follows the sequence of 1, 2, and 3 above.
5. All questions concerning the serving of females shall be decided upon by the establishment having in its care the female animal concerned.
6. All other questions concerning the future of the breeding herd shall be decided by a Board of Trustees upon which the following shall be represented:

Arizona Zoological Society	Shikar-Safari Club
Fauna Preservation Society	World Wildlife Fund
Zoological Society of London	

7. A studbook will be kept by the breeding establishment which will send the Fauna Preservation Society a copy and keep that society supplied with up-to-date information.

offered to meet the costs of constructing pens at Phoenix and of transporting the animals.

In January 1963 the FPS Arabian oryx committee accepted the Phoenix–Shikar offer, and the three Aden oryx were quarantined in New York, with the London female, before travelling on to the Phoenix Zoo, as it soon became on the death of Harold Maytag. At this point the World Herd existed, with four trustees: the Arizona Zoological Society (AZS), FPS, the Shikar-Safari Club and ZSL. The terms of the agreement concerning ownership and management of the animals were specified in detail (Table 3.1).

As the four oryx were not considered a secure or viable nucleus for a breeding herd, Grimwood searched for further captive Arabian oryx. HH Shaikh Jaber bin Abdullah al-Sabah, then ruler of Kuwait, had two females, both of which were offered to the FPS. One died in Kuwait but the other was quarantined in Mombasa, Kenya, before joining the herd at Phoenix in September 1963. A small herd of eight males and five females also existed in the private zoo of HM King Saud bin Abdul Aziz in Riyadh. After many approaches and some confusion, during which Arabian gazelles appeared to be under discussion, Grimwood finally obtained the gift of two pairs of oryx to the World Wildlife Fund (WWF), in exchange for some animals of Kenyan species. These four oryx were quarantined in Naples before continuing to Phoenix in July 1964, when the World Wildlife Fund (USA) joined the trustees. The World Herd thus had nine founder oryx.

3.4   Establishment of further herds in the USA

As the Phoenix herd gradually increased (3.5.3), a further herd of Arabian oryx was started at the Los Angeles Zoo. Its founders came from the Riyadh Zoo, and were assumed to be unrelated to the Saudi oryx which went to Phoenix. Riyadh Zoo was persuaded by a dealer to part with a male and one of only two breeding females, which were then sold to Los Angeles Zoo. This met with disapproval from the FPS and Species Survival Commission of the IUCN who felt that trading such rare animals for very high prices would stimulate further attempts to catch animals in the wild (Fitter, 1982). In any event, the female destined for Los Angeles gave birth in quarantine to a female calf by a different male, thereby adding to the genetic diversity of this herd. Although not belonging to the World Herd, the animals of this herd participated in the studbook records.

The trustees had envisaged that when numbers permitted, World Herd oryx should be dispersed to start further herds, principally to reduce the risk of catastrophic loss through disease. At the end of 1971 the herd numbered 30, and the trustees agreed that six animals of sex ratio 4:2 (male:female) should be transferred in 1972 to the San Diego Wild Animal Park (SDWAP) in California. Five of these oryx belonged to FPS under the ownership agreement (Table 3.1), the remaining one to WWF. The San Diego Zoological Society (SDZ) joined the trustees.

At the end of 1975 the Phoenix herd had increased to 45 oryx. In the following year, two males and two females were sent to Gladys Porter Zoo (GPZ) in Texas, and a further 11 oryx (ratio 4:7) were sent to SDWAP. At this stage the AZS unilaterally regarded all remaining oryx in Phoenix as its property, to be managed and disposed of as it wished.

3.5   Breeding in the World Herd and dispersals
3.5.1   *Longevity and fecundity*
Under equable, captive conditions, the Arabian oryx is long lived. The World Herd founders were all adult, though of unknown age, on arrival at Phoenix. Apart from one female dying after 1.8 years there, the other eight oryx lived in captivity for between 7.4 and 15.8 years. Life-spans of the first 15 born at Phoenix, excluding those dying perinatally or transferred out of the USA, ranged between 4.4 and 16.5 years, with more than 50% living to be more than 10 years old (studbook records).

Data in the studbook show that some females at SDWAP produced up to 14 calves throughout their life-spans (Table 3.2). More than half these females had their first calf before 30 months, while many calved before 24 months, requiring conception at 15 months (see below). Differences in management or breeding policy probably explain the greater age at first calving of the five females bred at Phoenix and later transferred to SDWAP.

The most common interval between successive calves was 8.1–9.0 months, followed by 9.1–10.0 months (Table 3.2). Together, these accounted for 64% of all intervals, with greater intervals progressively less common. As gestation length, based on two records at Phoenix, is 255–260 days, or 8.5 months (Turkowski & Mohney, 1971), the interval frequency shows that most oryx were re-conceiving on their *post-partum* oestrus. The Phoenix oryx had oestrous cycles of 25–32 days (Turkowski & Mohney, 1971). The longer intervals between the four Phoenix births in Table 3.2 was due to the practice of separating recently calved females from males (Dolan, 1976). Inter-calf intervals and seasonality in breeding are considered later (7.2).

3.5.2   *Mortality*
While mortality rates may be an important factor in overall population performance, it is difficult to calculate realistic figures for a captive population with many transfers if these take animals to institutions outside

Table 3.2. *Ages at first calf and death, total number of calves (including stillbirths) and distribution of inter-calf intervals of all females breeding at SDWAP 1973–84*

Female[a]	Age at first calf (months)	Months between calves[b]									Total calves	Age at death[c] (years)
		< 8.0	8.1 –9.0	9.1 –10.0	10.1 –11.0	11.1 –12.0	12.1 –13.0	13.1 –14.0	14.1 –15.0	> 15.1		
17*	40.4	1	3	6			1	1			14	15.3
30*	84.9			1							2	8.6
31*	81.0										1	7.4
37	28.5		1	1							3	4.4
39*	47.5		5	2	1	1					11	A
42*	59.4		1					1			3	11.6
74	22.7		5	3	1		3				12	A
78	22.9	(1)	7	1	1	1					12	A
82	23.0		5	3		1					10	A
93	29.5										1	A
96	23.9				1	1	2				5	T
112	21.3	(1)	6	1							9	A
113	28.2		2	1						1	6	9.1
120	23.9										2	3.1
123	36.4		1	1		1			1		5	A
125	23.8										1	T
137	29.6		2		1	2					6	A
138	29.2							1			3	T
139	27.5			1		1				1	3	T
148	25.8										2	T
153	33.5		2		2						5	A
175	32.3		2		2						5	A
409	25.4				2						3	A
433	27.8									1	2	A
549	24.1										1	N
550	21.9										1	A
Total born SDWAP only		3	42	21	11	8	6	3	1	3	98	

[a] *, born in Phoenix.

[b] / born in Phoenix; ( ) stillborn.

[c] A, alive at end 1984; T, transferred from SDWAP; N, no record in 1984 studbook.

the studbook recording scheme. Table 3.3 shows the age class mortality, the ages of animals when transferred out, and the standing age structure at 31 December 1985 of oryx born at SDWAP from the first calf in 1973 to the end of 1985. On this basis, live births totalled 139, of which 12.2% died within their first month. Mortality increased only to 15.1% for the first 6 months of life. Most transfers of oryx from SDWAP occurred at ages between 6 and 36 months, suggesting that survival was very high through this period or, at least, until departure from SDWAP. Mortality amongst older oryx remaining at SDWAP was also very low as far as 12 years of age. Therefore, captive oryx surviving their first month are then subject to low mortality, which confirms the figures on longevity, above, and the high proportion in Phoenix which survived to above 10 years old.

3.5.3   *Increase of the World Herd*

These reproductive attributes resulted in a rapid increase in World Herd strength. The growing numbers in each of the four main herds were recorded in the official studbooks (Table 3.4). Each grew rapidly before stabilising around a fairly steady level for each zoo, perhaps reflecting factors such as the amount of space available for the species and its changing conservation priority, the overall numbers in captivity, and the demand for animals for re-introduction or trade. These totals represent the balance between births, deaths and transfers in and out. The increase in captive numbers was due to the birth of 425 calves in these four collections between 1963 and 1985 (Table 3.5).

Table 3.3. *Oryx born at SDWAP 1973–85: frequency distributions of age at death or age at 31 December 1985 or age at transfer out of SDWAP*

Months			Years												Total
0– 0.9	1.0– 5.9	6.0– 12.0	1–2	2–3	3–4	4–5	5–6	6–7	7–8	8–9	9–10	10–11	11–12	>12	
Age Class															
Age at death															
17	4	0	0	2	1	0	0	0	0	0	0	1	0	0	25
Age at transfer out															
0	3	19	23	14	6	2	1	1	0	0	1	0	0	0	70
Age structure at 31 December 1985															
0	2	14	11	3	2	0	2	2	1	2	2	1	1	1	44

Table 3.4. *Oryx numbers in the four main herds from founding to 1984*

Herd	Year										
	1963	64	65	66	67	68	69	70	71	72	73
Phoenix	6	12	13	16	16	18	20	25	30	28	31
LAZ					3	5	7	8	10	11	15
SDWAP										6	7
GPZ											
Total	6	12	13	16	19	23	27	33	40	45	53
	1974	75	76	77	78	79	80	81	82	83	84
Phoenix	35	28	31	35	43	32	39	51	38	34	31
LAZ	17	18	21	20	23	30	17	12	10	13	10
SDWAP	10	13	27	41	42	49	46	46	36	45	42
GPZ			4	5	7	10	11	13	9	10	12
Total	62	74	84	105	123	127	108	102	88	96	95

LAZ, Los Angeles Zoo; SDWAP, San Diego Wild Animal Park; GPZ, Gladys Porter Zoo.

Table 3.5. *Number of calves born into four main herds from founding to 1985*

Herd	Year											
	1963	64	65	66	67	68	69	70	71	72	73	
Phoenix	1	2	2	3	3	3	3	5	6	6	7	
LAZ						2	2	1	2	2	5	
SDWAP												
GPZ												
Total	1	2	2	3	3	5	5	6	8	8	13	
	1974	75	76	77	78	79	80	81	82	83	84	85
Phoenix	6	8	8	8	13	12	8	8	11	15	10	14
LAZ	4	2	5	8	12	11	9	5	6	4	4	7
SDWAP	3	4	6	8	12	11	19	11	17	16	13	19
GPZ			1	2	3	5	5	5	3	3	5	
Total	13	14	19	25	39	37	41	29	39	38	30	45

Key to herds as in Table 3.4.

The total number of oryx in the three World Herd collections (Phoenix, SDWAP, GPZ) and Los Angeles Zoo are shown in Fig. 3.1 for the years 1963–85. From 1964, when all the founders were present at Phoenix, to 1978, when the first oryx were transferred outside this group of zoos, numbers increased exponentially with an annual 19% increase or a doubling of the population every 4 years (Fig. 3.1). From 1978 the total number of oryx reflects the yearly harvest to outside destinations, but for the period 1981–4 the four zoos between them maintained 90–100 oryx, yielding 30–40 calves annually. This last figure, less any mortality, was the number, of appropriate age composition, available for dispersal each year while a stable parent herd was maintained.

3.5.4   *Dispersals and exchanges*

As numbers built up, transfers between herds and dispersals became feasible without jeopardy to any group. These movements fall into three phases (Tables 3.6 and 3.7). As early as 1972, Los Angeles and Phoenix (AZS) exchanged oryx to diversify their genetic bases, and in 1972 and 1976 the San Diego and Brownesville, Texas, herds were started. A small herd at

Fig. 3.1. Oryx numbers in World Herd and Los Angeles Zoo 1963–85. ●, Numbers; ▲, log$_e$ numbers. The histogram shows the numbers exported from 1978. For 1978, (●) indicates actual population, (○) the number without any oryx exported.

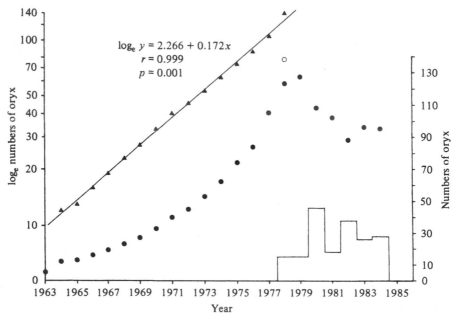

the San Diego Zoo was founded in 1978 with stock from SDWAP. At the end of 1976 the World Herd numbered 105 (Table 3.4), and the trustees agreed to start sending animals back to the Middle East in 1978. Between 1978 and 1980, 32 oryx went to four destinations (Table 3.7). Starting in 1979, three European zoos were provided with 16 oryx as founding stock, from which they were jointly committed to send their first 16 offspring to ZSL and FPS

Table 3.6. *Transfers of oryx from World Herd and other zoos to within USA*

Year	No.	To	From	No.	To	From
1972	1	AZS	LAZ	1	LAZ	AZS
	6	SDWAP	AZS			
1976	1	AZS	LAZ	1	LAZ	AZS
	8	SDWAP	AZS	4	GPZ	AZS
1977	6	SDWAP	LAZ			
1978	3	SDZ	SDWAP			
1979	12	SDWAP	AZS			
1980	2	SDZ	SDWAP	16	LAZ	IWP
	8	AZS	IWP/IAE	4	LAZ	SDWAP
1981	4	SCSC	SDWAP	6	IWP	LAZ
	4	IAE/IWP	AZS			
1982	4	IWP	LAZ	21	IAE	SDWAP
	5	IAE	AZS	4	LPZ	SDWAP
	1	IWP	SDZ	5	IWP	GPZ
	3	SCSC	GPZ	1	GPZ	IWP (b.LAZ)
	1	IWP	GPZ			
1983	8	TR	AZS	9	IAE	AZS
	2	IAE	SDWAP	2	IWP	GPZ
	2	SDWAP	SDZ			
1984	4	IAE	AZS	4	IAE	SDWAP
	5	IWP	LAZ	2	LD	SDWAP (loan)
	3	KZ	SDWAP (loan)	2	HI	SDWAP
	1	LAZ	SDWAP	2	ZZ	LAZ
	2	SZ	LAZ			

AZS, Arizona Zoological Society.
GPZ, Gladys Porter Zoo.
HI, Holiday Island.
IAE, International Animal Exchange.
IWP, International Wildlife Park.
KZ, Knoxville Zoo.
LAZ, Los Angeles Zoo.
LD, Living Desert.

LPZ, Lincoln Park Zoo.
SCSC, St Catherine's Survival Centre.
SDWAP, San Diego Wild Animal Park.
SDZ, San Diego Zoo.
SZ, Sacramento Zoo.
TR, Tasajillo Ranch.
ZZ, Zoo Zacango.
b., born.

in return for the animals donated by them to the founding World Herd. This manoeuvre circumvented the ban on hoofed stock entering Britain direct from North America.

The completion of these deliveries coincided with, or resulted in, changes in World Herd management (3.9), through which ownership of all oryx reverted to the zoos holding them. As numbers in these four primary sources again rose, dispersals were increasingly made to the International Wildlife Park and International Animal Exchange. These acted as secondary sources of oryx for further new herds and for export, as to Saudi Arabia in 1982 (Table 3.7). Through the 1980s the primary sources also supplied oryx for new herds in the

Table 3.7. *Transfers of oryx from World Herd and zoos in USA to Europe, Middle East and Morocco*

Year	No.	To	From	No.	To	From
Middle East and Morocco						
1978	4	Jordan	SDWAP	8	Israel	LAZ
1979	1	Jordan	SDWAP	3	Jordan	AZS
	3	Morocco	SDWAP (1 b.AZS)			
1980	8	Oman	SDWAP	3	Israel	SDWAP (3 b.AZS)
	2	Oman	GPZ			
1981	4	Oman	SDWAP (1 b.AZS)			
1982	29	Saudi Arabia	IWP (1 b.SDZ 2 b.GPZ 7 b.AZS 5 b.LAZ 14 b.SDWAP)			
1983	2	Abu Dhabi	IAE (b.AZS)	2	Abu Dhabi	IAE (b.SDWAP)
	3	Oman	SDWAP			
1984	2	Dubai	SDWAP			
Europe						
1979	4	TP Berlin	AZS	1	TP Berlin	SDWAP
	4	Zurich	AZS			
1980	5	Rotterdam	AZS	2	Zurich	SDWAP (3 b.AZS)

Key to zoos as in Table 3.6.

USA, while the three European zoos, Zurich, Tierpark Berlin and Rotterdam, provided oryx to Hanover and Marwell Zoos, as well as London.

Thus, by the mid 1980s the Arabian oryx had multiplied and gone forth to many different collections. An approximate, but certainly an underestimate of world numbers is given in Table 3.8. The 172 oryx in the US were held in 13 known collections, while the European herds numbered six, and 10 countries in the Middle East, with Morocco, had oryx. The number of oryx in the latter group was almost three times that of the US population and, perhaps more significantly, many were of unknown but different genetic origin from the oryx in the Western world. However, out of over 700 oryx, only 34 (5%) were living wild in unenclosed natural habitat in Oman.

3.6 Breeding policies in captivity

While the increase in captive oryx numbers clearly ensured the survival of the species in the Western world, the long-term future of the

Table 3.8. *Arabian oryx status in different regions, 1984–6*

USA			Europe			Middle East and Morocco		
Location	No.	Source	Location	No.	Source	Location	No.	Source
SDWAP	42	a	Zurich	10	a	Abu Dhabi	133	b
Phoenix	33	a	TP Berlin	8	a	Qatar	100	c
IWP	33	a	Rotterdam	4	a	S. Arabia	70+	c
GPZ	12	a	Hanover	3	a	Bahrain	60	c
LAZ	10	a	London	3	a	Jordan	70	d
SCSC	10	a	Marwell	2	a	Oman	34	e
TR	10	a				Israel	29	f
SDZ	6	a				Dubai	15	c
LPZ	6	a				Kuwait	10	c
KZ	4	a				Morocco	7	a
ZZ	2	a				Iraq	?	
LD	2	a						
HI	2	a						
Totals								
Oryx:	172			30			528	
Herds or countries:	13			6			10	

Key to USA zoos as in Table 3.6.
Sources: a, 1984 Studbook; b, J. C. M. Lewis (personal communication 1986); c, D. M. Jones (personal communication 1986); d, M. Abu Jaffar, unpublished report to IUCN (1986); e, this work; f, W. Clark (personal communication 1986)

species in captivity and the viability of re-introduced animals depend on the genetic status of the population (Soulé *et al.*, 1986). Specifically, if their animals are ever to be restored successfully to the wild, captive populations should be managed for maintenance of as much genetic diversity as possible through maximising effective population size and attempting to equalise contributions from the founders (Foose *et al.*, 1986; Frankham *et al.*, 1986). This may require active exchanges between herds and the rotation of breeding males. The following data from official studbooks provide some insight into the practices of those zoos with World Herd oryx.

From 1963, when Phoenix Zoo had the original nine founders and the first three calves, it drew up a breeding programme with the survival of the species and minimisation of inbreeding as the foremost considerations (Thorp, 1964). The 12 animals were divided into Grimwood (i.e. Operation Oryx) and Saudi groups, reflecting their distinct origins and parentage. The plan was to breed within each group for three years, followed by intergroup breeding for one year. Despite the death in 1965 of the Kuwait founding female, without any offspring, the calves born over the next few years suggest that this plan was implemented.

At Phoenix between 1963 and 1986, a total of 167 calves were produced by 22 males (Table 3.9). Three wild-born males produced 31 calves and a single immigrant male from LAZ sired 20 calves. The picture is of frequent changes of breeding males, which resulted in no father–daughter matings.

The pattern at LAZ is different (Table 3.9). Here, four males sired 96 calves, but all but four of them were offspring of one wild male (42 calves) and one from Phoenix (50 calves). The long breeding periods of these two males resulted in 14 father–daughter matings. The SDWAP pattern seems similar in that five males produced between them 153 calves, except that all the males originated from other herds – three from AZS and two from LAZ. Sixteen of the calves resulted from father–daughter matings. Male 67 bred for so long that he sired two calves by a female who was his daughter and grand-daughter at the same time. In GPZ, paternity of the 31 calves born is more evenly spread between three males, two from AZS and one from LAZ, although one calf resulted from a father–daughter mating.

The rapid increase in oryx numbers at Phoenix and Los Angeles in the 1960s suggest that the dire effects of a genetic bottleneck were successfully avoided by rapid expansion of the population from its founding level (Foose, 1983). Initially, Phoenix had no scope to import unrelated breeding males, yet its breeding programme certainly prevented father–daughter matings, and probably avoided an increase in the level of inbreeding. The zoo's facilities, its intense concern for the species in the early years of the World Herd (Thorp,

1964; Turkowski & Mohney, 1971) and the relatively small number of animals involved must all have promoted efficient execution of the breeding plan. By comparison, the bulk of SDWAP's Arabian oryx occupy a single large enclosure in which animal exchange is less easily accomplished. This suggests that, although males from other zoos were used, the steady production of calves was adequate reason for not rotating the males. The production of all 123 calves at SDWAP over 11 years (Table 3.2) by 26

Table 3.9. *The number of calves from each breeding male, his origin, and the number of father–daughter matings in four USA herds from founding to early 1986*

Phoenix Zoo

Male studbook no.	2	3	6	7	10	12	16	38	33	40	45
No. calves	11	13	1	8	7	14	10	2	2	3	9
No. father–daughter											
Male origin if from outside	Wild	Wild			?Wild						

Male studbook no.	49	52	68	88	89	132	126	165	168	173	460
No. calves	12	10	20	1	10	7	1	9	2	8	2
No. father–daughter											
Male origin if from outside			LAZ								

San Diego Wild Animal Park

Male studbook no.	21	35	12	67	65
No. calves	9	33	5	93	13
No. father–daughter		5		11	
Male origin if from outside	AZS	AZS	AZS	LAZ	LAZ

Los Angeles Zoo

Male studbook no.	59	29	79	80
No. calves	42	50	3	1
No. father–daughter	10	4		
Male origin if from outside	Wild	AZS		

Gladys Porter Zoo

Male studbook no.	28	85	155
No. calves	13	13	5
No. father–daughter		1	
Male origin if from outside	AZS	AZS	LAZ

Abbreviations for zoos as in Table 3.6.

females, with short intervals between calves, also indicates that management was directed to high productivity. In this case, the level of inbreeding may have risen.

Breeding the Arabian oryx in captivity has enjoyed several advantages. The genetically effective founders were seven oryx at Phoenix and two at Los Angeles. As these were probably all unrelated and non-inbred, they would have contained over 90% of the available genetic diversity, measured as heterozygosity. The proportion would have declined drastically with any fewer founders (Foose, 1983). The early breeding policies paid dividends, for founder representation was fairly equal in the oryx of the 1984 studbook (Fig. 3.2). A large proportion of the original genetic diversity is also preserved, for a genedrop analysis (Mace, 1986) showed that only two of the nine founders had lost more than 5% of their genome in the 1984 population. This was due entirely to the early breeding programme, which resulted in the founders having many offspring, many of which also bred.

The oryx population rapidly reached its carrying capacity in the main zoos, and the increasing dispersals to new or other collections had demographic consequences. Comparing the population during the growth years of 1972–6 with the larger population of 1980–4, the annual, finite rate of increase decreased from 118% to 114% and the estimated generation time decreased from 8.8 to 7.7 years (Mace, 1986). These combine to cause the minimum

Fig. 3.2. Percentage representation of major founders of the World Herd in the living population of the 1984 studbook (from Mace, 1986).

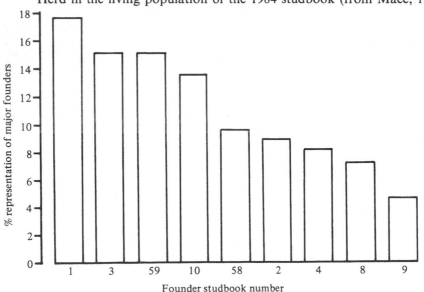

effective population size, which will safeguard 90% heterozygosity for 200 years, to increase from around 500 to around 2000 individuals. In other species (Ryman *et al.*, 1981), effective population size is especially sensitive to the removal of animals in the one to two year age class, the main group being transferred from SDWAP (Table 3.3), and then often lost to the studbook.

Reliance on a few successful breeders in latter years, and particularly on males, must have resulted in increased inbreeding in the captive herd. Until 1972, there was no inbreeding in the World Herd, a tribute to early management policies (Mace, 1988). Since then inbreeding has increased and 43% of those in the 1984 studbook were inbred to some extent, with 15% having coefficients greater than the 0.125 level of first-cousin matings (Mace, 1986). Although this level may be tolerable in zoo populations, the relationship between inbreeding level and calf survival in Oman (7.5) indicates that, if average inbreeding in the captive population rose to this level, zoo oryx might not be viable for re-introduction.

3.7   Planning the re-introduction to Oman
3.7.1   *Feasibility studies*
The extermination in Oman in 1972 of the world's last herd of wild oryx went largely unnoticed within the country, although reported outside (Henderson, 1974). However, many Harasis said in 1986 that they knew at the time that the last herd had been eliminated, and were angry and sad that their sons would never know the oryx.

Only two years after this event, the first steps towards the eventual re-introduction of the oryx resulted from a meeting between the Ruler of Oman, HM Sultan Qaboos bin Said and his newly appointed adviser for conservation and development of the environment, R. H. Daly. This audience resulted in Daly being commanded to find how the oryx might be restored to Oman as part of the country's heritage and, thereby, fulfill the second and more ambitious aim of Operation Oryx and the World Herd (Daly, 1985). Already, the potential of a re-introduction project to provide worthwhile and long-term employment to desert-dwellers was appreciated. It would be an attractive alternative to their becoming unskilled labour in oil camps for the sake of a wage, as was feared would happen as oil exploration spread throughout Oman (Thesiger, 1959).

The feasibility of a re-introduction was assessed by Dr H. Jungius, as consultant from the WWF. With Daly and a veterinarian, Dr M. H. Woodford, who had filled the same role on Operation Oryx, he travelled extensively through the interior of Oman in February 1977 and March 1978.

They were joined on the second survey by a botanist, Dr R. M. Lawton. Each produced an unpublished report, summarised by Jungius (1985).

The first survey concluded that the Jiddat-al-Harasis was suitable oryx habitat, and, indeed, little ecological change since the last oryx lived on it only five years previously could be adduced (Jungius, 1977). The 1978 survey concentrated on the Jiddat-al-Harasis and selected an area known as Yalooni for the project site (Fig. 4.1). It was chosen as the best vegetated pan (4.2) on the Jiddah, with resources of grazing, shrub and tree browse, and because of its location far from existing roads. A water-well was conveniently situated within reasonable reach. It was also recommended that the whole Jiddat-al-Harasis should be proclaimed a wildlife reserve or sanctuary with a law-enforcement capability, before any oryx were taken to the area (Jungius, 1978b).

Jungius (1978b) recommended that the World Herd should provide a minimum of 12 oryx, in a mixture of both sexes and a range of ages, which should be acclimatised in a 1 km² enclosure at Yalooni. This nucleus would provide a breeding herd, from which groups of optimal composition could be released into the desert when ready, and at opportune times, without depleting the parent breeding-stock. This was a conservative approach, reducing the risk that the entire stock of founders might disappear at once on release. On the other hand, this dictated a spell of breeding in the enclosure which would prolong the period between the arrival of oryx in Oman and the first release.

3.7.2   *Project design*

With the optimistic conclusions of the feasibility studies, Daly drew up financial estimates for the re-introduction, and designed the facilities and accommodation, for both oryx and people, to carry out the project along the recommended lines. The remoteness of the project area required a self-contained camp with full services, a workshop and communications equipment, to which all fuel, water and food would have to be trucked through the desert, with no graded road within 40 km. A staff establishment was drawn up to retain as much flexibility as possible to cope with any eventualities arising once the oryx were in Oman. These aspects are described in further detail in Chapters 4, 5 and 8. The oryx facilities are described in Chapter 5.

In late 1978 the plans and estimates were complete for what had become the Jiddat-al-Harasis Development Programme, Part B. Part A was the much larger development of a Tribal Administration Centre, 80 km from Yalooni

(4.8.1), which was completed in 1981. HM Sultan Qaboos approved the re-introduction plans in late 1978, directing that funds should be available for 5 years from January 1979. Construction of the camp started in September 1979, and by early 1980, when the first oryx were scheduled to arrive, the main oryx enclosure, pens, staff accommodation and services were in place.

Since 1974, the Office of the Adviser for Conservation of the Environment has been within the Diwan of His Majesty The Sultan for Protocol, subsequently called the Diwan of Royal Court, in the capital, Muscat. By virtue of this position, the re-introduction project was always part of the Diwan rather than any government ministry, although it would not have fitted into the remit of any particularly ministry. In each year, the project manager submits a budget estimate to the Diwan for approval by the Ministry of Finance. In the interests of efficient administration and prompt action, despite Yalooni and Muscat being 500 km apart, the project is permitted to operate with a bank account rather than having to follow more cumbersome, government procedures of administration, purchasing and payment. However, all personnel and salary matters are handled by the Diwan. Since it started, the environment adviser has directed the project, with a biologist field manager at Yalooni reporting to him and responsible for day-to-day running of the project and its technical aspects. The project's position within the Diwan, the concessions for its administration and its simple managerial structure are all relevant to progress on the re-introduction (8.2).

3.7.3 *Oryx for Oman*

Preliminary contact with the trustees of the World Herd in 1978 indicated that they would be likely to provide, at no cost, oryx for Oman's re-introduction. The provision of 12 oryx by 1980 was a priority agreed at a meeting in February 1978. SDWAP undertook to assemble the animals either from its herd or others.

Two important problems had to be resolved at this stage. First, Oman's proposal involved 12 animals, with equal numbers of males and females but with a range of ages, preferably even including a female who had bred. The oryx initially selected by SDWAP were all from its 1978 calf crop, and reassurance had to be sought that at the time of shipment the group would contain a wider range of ages. Second, in the period since the feasibility study, the strategy for releasing oryx changed. It was realised that oryx social organisation and the facilities at Yalooni would not allow a breeding herd to be maintained in the enclosure while small subgroups were released when ready. It would be essential to develop a herd which would mimic an ancestral, natural herd as far as was possible, which would then be released as a unit. This 'integrated herd' would include all 12 founders, leaving no residual breeding herd. For this reason, also, it was essential to have a range of ages in the founders, and it would speed up the process of release in Oman if the 12 animals had been together in the USA. This would allow a stable dominance hierarchy at least to start developing, and if any individuals were clearly unsuitable, for whatever reason, they could be replaced. Once in Oman, no substitutions would be feasible. However, despite many requests from Oman and IUCN, and assurance from SDWAP that the 12 oryx should run together in the USA, it transpired that SDWAP had no spare animal accommodation for this purpose. Thus, by late 1979 it was understood that the herd would have to be integrated in Oman.

Integration of the first herd in Oman was prolonged because the 12 immigrants came in three successive shipments in 1980 and 1981 (5.2.3). The last group should only have numbered two oryx, but an extra two males were sent from SDWAP, following requests for a further supply of oryx to Oman. As these oryx could not be integrated into the first herd before its scheduled release date, they were earmarked for a second herd for release, despite no assurance of any further oryx from the USA. Subsequently, WWF International requested three females from SDWAP for development of a second herd following a visit to the re-introduction project by its president, HRH The Duke of Edinburgh, after the first herd's release into the desert. When these females had arrived in Oman in 1983, the re-introduction project had received a total of 17 oryx from World Herd sources.

3.7.4   *Wildlife protection measures in Oman*

In view of the causes of the original oryx's demise (2.6), their re-introduction was doomed unless protection of the species was both proclaimed and enforced. Modern conservation measures in Oman started in 1964 when Sultan Said bin Taimur decreed that vehicles should not be used for hunting gazelles and oryx. None the less, military parties subsequently all but eliminated the Arabian gazelle from western Oman, and the last oryx there disappeared in about 1970 (2.5). At the same time Sultan Said commanded four Harasis be employed to protect their gazelles, and the tribe was probably the only one to continue its traditional hunting methods when car ownership started in the 1970s (4.8.2). In 1980 these guards were expanded to 14, who operated as Royal Oman Police 'civil guards'. Their two patrols ranged widely over the Jiddah with prime responsibility for gazelle protection.

In 1976 a Diwan Ministerial decision protected from hunting, capture or interference all large mammals, which were specified and included the oryx, together with the houbara bustard, the favourite prey of falconers. All marine birds were totally protected, but a limited shooting and hawking season for Oman's four species of sandgrouse was defined. Although exemptions to any restrictions could be obtained only with HM The Sultan's permission, he stated unequivocally in late 1977 that no permission would be given to foreign hunting parties to enter and operate in Oman. Border security has ensured that any infringements have been only minor.

These measures ensured the protection of released oryx, and before the release of the first herd in 1982 the attention of oil companies was drawn to the 1976 regulations. In the same year, an arbitrary oryx project area was declared on the basis of the likely limits of oryx movements in the short term (Fig. 4.6). Any companies operating in this area were obliged to liaise with the conservation adviser in Muscat and with the oryx management to avoid unnecessary environmental damage and the possibility of oryx being driven to flight or being harmed by their operations. The same area was gradually recognised as a military no-go area except by prior arrangement with the adviser or in the national interest.

Although these measures effectively ensured the safety of the oryx, the re-introduction project had no formal status as a protected area, and local land-use conflicts had to be resolved on the basis of mutual interests and benefits (8.4). A comprehensive protected area system in Oman has been proposed (IUCN, 1986), in which 27 500 km² of the Jiddat-al-Harasis would become a national nature reserve. Development of this system will also include comprehensive legislation for wildlife conservation.

3.8 Other oryx re-introductions in the Middle East

Oryx from Los Angeles Zoo were sent in 1978 to Israel to establish a captive herd in the Negev desert (Table 3.7), with release into a large reserve the ultimate aim.

In 1978–9, oryx from the World Herd travelled to the Shaumari Wildlife Reserve in Jordan. The reserve is an area of 22 km² originally fenced in 1958 for an experimental study of desert farming. In 1975 the Royal Society for the Conservation of Nature (RSCN) restored the fence to be proof against livestock, and the vegetation soon showed a dramatic improvement inside, augmented with some exotic tree species.

A feasibility study confirmed the area as suitable for oryx (Clarke, 1978). The founders were eight oryx from the World Herd and three from Qatar in 1978–9. Initially, the oryx occupied a 3 km² enclosure in which heavy supplementary feeding was required. In late 1983 the herd was released into 22 km² where the perimeter fence had been renewed with financial assistance from Oman. Here, the well-developed vegetation and regular winter rain provided adequate grazing from 1983 to 1986, and piped water was freely available. One male was transferred to Oman in 1984, and in the same year three males were received from Zurich Zoo. Oryx numbers increased from 11 at the end of 1978 to 70 at the end of 1986 (Fig. 3.3). This is a 28% increase per year, with the population doubling every 2.75 years. The increase was partly due to a death rate of only 11 oryx over this period. The success of this herd has led to plans to increase the protected area to 100 km².

3.9   Dissolution of the World Herd and trustees

The trustees held their fourth meeting in November 1979. According to the minutes, oryx in the three World Herd collections and Los Angeles then numbered 137. World Herd oryx had been sent to Jordan, Berlin, Rotterdam and Zurich, while the first shipment to Oman was a few months away. After reviewing the status of all known captive herds of oryx, the SDWAP representative said his institution intended to keep a herd of 40–50 females 'against catastrophe and/or failure of re-introduction', and that a viable herd would be maintained solely for free distribution of oryx into the wild. This was unanimously approved and the generosity appreciated. The next motion ensured that the European zoos with oryx should send their

Fig. 3.3. Oryx numbers in Shaumari Reserve, Jordan, 1978–86. (●) Numbers; (▲) $\log_e$ numbers. Point for 1978 excluded from regression.

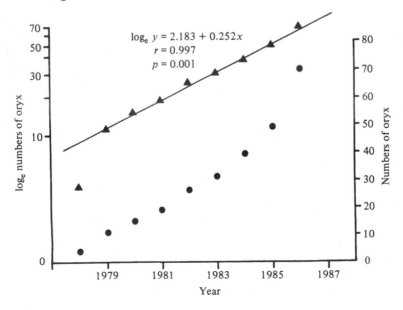

$$\log_e y = 2.183 + 0.252x$$
$$r = 0.997$$
$$p = 0.001$$

offspring to London Zoo until the collective obligation (3.5.4) was discharged, after which ownership of the animals reverted to the zoos. Three further motions granted ownership of their oryx to any UK zoo, the GPZ and the SDZ, but it was agreed that disposal of offspring from these collections should be as directed by the World Herd trustees with the advice of the Captive Breeding Specialist Group (CBSG) of the IUCN Species Survival Commission.

The trustees then concluded that these resolutions had, in effect, relieved them of any trusteeship functions. After a motion that a special subcommittee of the Antelope Specialist Group should be created to oversee management of the Arabian oryx, the trustees proposed their dissolution on 31 March 1980, subject to the assent of their various governing bodies. This was duly given.

The Antelope Specialist Group never took over this new responsibility for the Arabian oryx, leaving the CBSG in an advisory role. Direct communication between Oman and SDWAP proved fruitless and, despite a request for CBSG's assistance in 1984 to secure more animals for the re-introduction, dispersals from the former World Herd institutions continued (Table 3.6). In 1986, the CBSG advised against issuing an export permit from the USA for oryx in the light of Oman's outstanding request. This stimulated an examination of the status of the North American oryx population (Mace, 1986), with the CBSG soon urging that a Species Survival Plan should be developed to the American Association of Zoological Parks and Aquaria format (Conway, Foose & Wagner, 1984). The plan would define the requirements for a captive population that would be secure genetically and demographically. A further aim of the prospective plan would be provision of 10 oryx per year over the first 10 years for re-introduction projects (9.5.1).

3.10 Discussion

The rapid increase in oryx numbers in the four main herds confirmed early conclusions that the species is easy to keep in captivity (Thorp, 1964; Turkowski & Mohney, 1971). Captive breeding of oryx species and other hippotragine antelopes has been so successful that secure populations of all exist in north America (Thomas *et al.*, 1986). The increase in World Herd oryx was due to a combination of early maturity, high fecundity, as measured by the inter-calf interval, and a high survival rate of calves over 6 months of age. This occurred under captive conditions in an exotic climate at latitudes 30°–35° N, with water provided *ad libitum*, and an adequate level of feeding with little seasonal variation. The data from Jordan show that under a native climatic regime at latitude 32° N, with water provided and whether being fed or grazing, a greater rate of increase was possible over a short run of years. In common with other large antelopes (Thomas *et al.*, 1986), the Arabian oryx has simple requirements in captivity, and the species is clearly highly productive if its minimum requirements are met under a wide range of different conditions.

The north American population derives its present strength from creation of the World Herd. This, in turn, depended on oryx from Operation Oryx, which was also the stimulus to locate and draw together further captive animals to form a viable breeding nucleus. Mounting the capture operation was a bold risk, and it probably took place in the last year with any prospect of success. Its sponsors, the FPS, can claim a major share of the credit for the species' present status. If the World Herd had not existed, it is doubtful if the Los Angeles herd, with three founders, could have thrived to the same extent, for it would have started genetically depauperate, and inbreeding would have been unavoidable.

In the early 1960s, information about the mere existence of captive oryx herds in many parts of Arabia was hard to obtain, and a complete and accurate inventory of herd numbers, locations and sizes is still not feasible. It is likely that there were always, and still are, more oryx in the Middle East than are acknowledged. The example of Qatar's periodically large captive numbers shows that these herds persisted, and would probably have continued to do so without the marginal additions from the USA. On this basis, the World Herd cannot be attributed with saving the species from extinction.

The level of management and record-keeping varies in the Middle East between herds with meticulous standards and others in which individual oryx are not identifiable, and there are, therefore, no accurate data on animal history or breeding performance. Some collections have been known to possess hybrids between Arabian and other oryx species. Periodic outbreaks of decimating diseases occur in some captive herds (Fitter, 1982; Woodford, 1989). For these reasons, and because many Middle Eastern herds are privately owned without obligations to donate or transfer oryx between sites, these herds are not prime sources of animals for re-introduction for the present. For projects such as Oman's re-introduction, oryx from USA herds with ancestries that can be traced to the founders and are of known relationship to one another, and to animals already in Oman, are the best stock. The many advantages of known pedigrees will be amply demonstrated in subsequent chapters, and they outweigh the cost, logistic and animal health implications of transporting oryx between continents.

Efforts to re-introduce the oryx to the wild have highlighted the importance of the relationship between the producers of animals, the captive-breeders, and the users. This has a further significance in that the time between creation of the World Herd and the first release back into the wild was only 20 years, and 10 years after the last wild oryx were killed. Captive management plans have adopted the standard aim of maintaining 90 % of the founders' genetic diversity for 200 years (Soulé *et al.*, 1986). The time period is arbitrary, but is an estimate of the probable, average duration of captive-breeding before re-introduction is feasible. Thus, history may show the Arabian oryx re-introduction to have followed an atypically short captive phase.

When any species' captive-breeding programme is formalised, it must contain comprehensive details on the terms and conditions of any future re-introduction, and the obligations on the captive-breeders. These will vary with the species, depending on many factors including the origins of the captive population. The Arabian oryx World Herd constitution enshrined re-introduction as a primary aim but, obviously, no time scale could be envisaged. Consequently, there was no detail of the extent of the World Herd's obligation to provide animals, free upon request, for re-introduction. The trustees evidently felt that agreeing to provide eight oryx to Jordan and 12 to Oman met this obligation. These requests were made at a time when the World Herd was large enough for animals to be dispersed, and the trustees might have been able to decide on an appropriate balance between the dispersal of animals for re-introduction and for sale or exchange with other zoos. Their dissolution at this time, soon to be followed by large dispersals of oryx to start other captive herds, suggests that commercial interests prevailed. Despite the generous SDWAP offer to keep a herd against re-introduction catastrophe (3.9), the herd never reached its promised size, and this institution was never subject to formal obligation to provide oryx in the years ahead. Oman did receive five oryx from SDWAP in the period after dissolution of the trustees, but the course of action for any future re-introduction project is, hopefully, clearer with the adoption of the Species Survival Plan as a collaborative management plan for captive herds (9.5.1). The history of the World Herd emphasises the need for specific guidelines concerning conservation priorities for the oryx, whether in captivity or back in the wild, with unified and co-operative genetic and demographic management. There is also a role for a clearly defined body to which the re-introduction project can communicate its needs, and which includes the captive-breeders. Its decisions should be binding, as it establishes the right balance between maintaining animals for re-introduction at some cost to their institutions and their using the species as a resource to be traded.

4

The re-introduction area and the Harasis tribe

4.1   Introduction

The aim of this chapter is to define the setting of the oryx re-introduction project and to describe it in terms of its infrastructure, climate, vegetation and wildlife, and the ecology of its pastoral tribe. As the people are an integral component of the project, their way of life both before and after the advent of modern Oman is described in some detail where relevant to the oryx project.

A Jiddah is a stony plain, so the Jiddat-al-Harasis is firmly labelled as the homeland of the Harasis people. It is a discrete geological and geomorphological unit, and much of the following description is due to the unpublished observations of M. W. Hughes Clarke. The characteristic plant communities of the Jiddah develop because of the area's geology and its unusual climate through the influence of the nearby Arabian Sea. Consequently, there has evolved an ecosystem of surprising productivity and diversity of animal life.

4.2   Geology and geomorphology

The Sultanate of Oman lies at the south-eastern extremity of the Arabian peninsula (Fig. 4.1), with a total land area of 212 500 km². Most of the country lies at latitudes lower than the Tropic of Cancer (23° N), thus falling within the vast belt of desert stretching north of the equator from west Africa through Arabia and into Asia.

The main determinants of Oman's surface geomorphology are the two ranges of mountains; the Hajar range in the north, and the Dhofar mountains in the south, rising to 3000 and 1920 m, respectively (Fig. 4.1). A complex system of water sources (wadis) drains the inland face of both. The wadis from the Dhofar range run north until deflected and overrun by the sand-dunes of the Rub' al Khali. Those from the Hajar mountains run south-west across a gently sloping pediment covered in gravels formed by water erosion after the range's uplift in late Eocene and Oligocene (30–40 million years BP). These wadis dissipate into the Umm as Samim *playa*, the northern end of the

Fig. 4.1. Oman's location in Arabia and general geomorphology. Line A···A is from Glennie (1977). (-----) International boundaries.

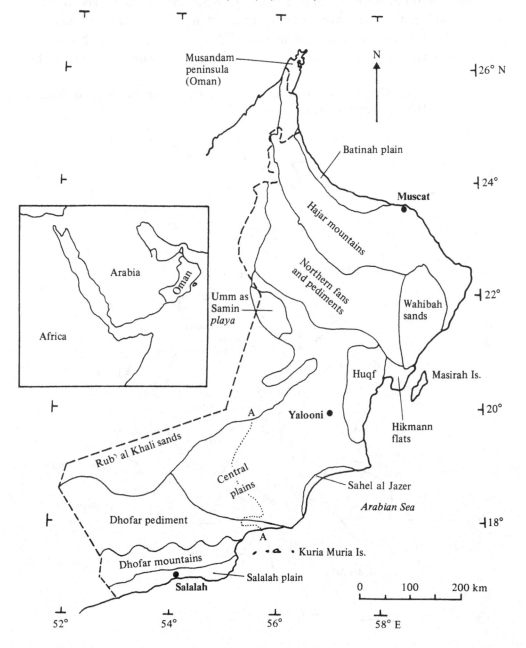

Huqf depression or into the coastal sand flats of the Hikmann area. Between these two drainage systems lie the central plains, without any pronounced surface water courses.

The western portion of these plains is composed of the same Eocene–Palaeocene limestones as the Dhofar range, but around 56° E this is overlain by a Miocene limestone (line A···A, Fig. 4.1). Its fossil content of Foraminifera dates deposition at some 13 million years BP when global sea-levels were up to 200 m higher than they are today (Harland *et al.* 1982). The fossil mollusc fauna indicates shallow marine conditions at its eastern border, becoming increasingly lagoonal to the west. The resulting contemporary plateau is the

Fig. 4.2. Geomorphological and ecological zones in and around Jiddat-al-Harasis, central Oman.

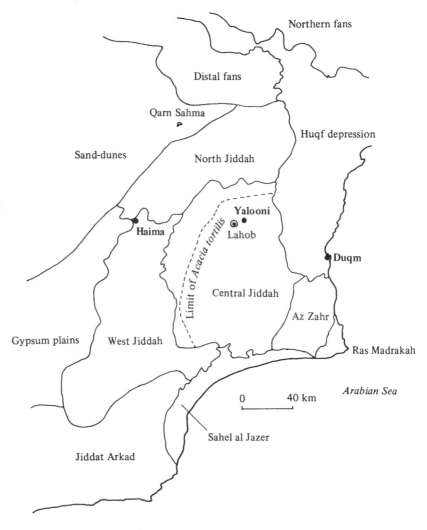

Jiddat-al-Harasis, which shows few structural features to indicate warping or uplift after deposition, and has an altitude of 100–150 m above sea-level.

The Jiddah is a distinct unit (Fig. 4.2) on both geomorphological and vegetational criteria. The same parallels occur in its three main subdivisions. The largest zone is the central Jiddah, bounded to the east by a 100 m escarpment, dropping to the Huqf depression, a highly saline alluvial plain with discontinuous outcrops of sedimentary rocks of almost every age from pre-Cambrian. The Huqf escarpment is abrupt, but to the south-west of Duqm the Jiddah limestone ends with a lesser gradient over the Oligocene limestone of Az Zahr, a low coastal plateau gently inclined to the south-east. Ras Madrakah is an ophiolite headland, rising to 205 m. The southern face of the Jiddah is separated from the Arabian Sea by a coastal plain of sand and recent sediments, the Sahel al Jazer. The wadis along this face are deeply dissected through both Miocene and Oligocene limestones.

In the south-west, the Jiddah abuts the Jiddat Arkad, whose rock is Oligocene limestone, with a well-developed dendritic drainage system running into the Sahel al Jazer. The western boundary of the Jiddah is less distinct but marked by a gradual transition into rolling plains with a gypsiferous caliche sediment 5–10 m thick overlying the Miocene limestones. This causes a major change in vegetation cover (4.5). The gypsiferous zone extends westwards to the southern pediments (Fig. 4.1). To the north-west, the Jiddah is limited by the southernmost sand-dunes of the Rub' al Khali, which overlie a very flat plain of Miocene limestone without any drainage features. The dunes extend eastwards to enclose a salt-plug, Qarn Sahma. The Jiddah extends northwards until limited by the head of the Wadi Huqf. This wadi has cut into the level Miocene limestone surface at its junction with the base of the northern pediments slope. The latter's surface rises at 1.25 m/km north-eastwards from this point to the Hajar mountain range over 200 km distant. The north Jiddah sub-division is distinguished from the distal fan zone by its vegetation cover (4.5).

The Jiddah surface is very flat, with few low ridges and outcrops exceeding 10 m, but surface drainage patterns are the basis for identifying three subdivisions (Fig. 4.2). Satellite imagery of the surface of the central Jiddah shows poorly developed dendritic drainage systems that are barely apparent on the ground. Their development started under more pluvial conditions in the Pleistocene (Beydoun, 1980). The Lahob feature is a double concentric, meteorite impact crater with an outer diameter of 6 km. Seismic sections show a central toroid crater several hundred metres deep, which has since been infilled by younger sediment, probably by wind action. This and further scars of likely meteorite impacts on the Jiddah show that the Miocene land surface has changed little since deposition. The rock surface is highly resistant to weathering, and the Jiddah is an area of neither deposition nor erosion.

On the ground, the most conspicuous features of the central Jiddah are scattered depressions or pans, typical solution hollows of desert limestone areas (Goudie & Wilkinson, 1977), known locally as *haylahs*. Their areas vary between 1 and 25 ha, with a few larger, and they support most vegetation (4.5). Between the sand-filled *haylahs*, the surface is a *hammada*, with rock cover varying between coarse sand and pebble surfaces, fields of loose, irregular limestone blocks rarely more than 20 cm in any dimension, and smooth stony pavements with highly weathered and closely fitting limestone flakes. All such stones develop a desert varnish on their exposed surfaces, often with profuse lichen colonies.

Hammada surfaces are probably formed by wind removal of fine material. With the prevailing climate (4.4), little sand accumulates on the surface of the Jiddah, and the ridges comprise exposed bedrock. Under a pavement surface there is no more than 15–20 cm of sand and limestone fragments above the bedrock. Coarse, unstructured sand accumulates only in hollows, where it may reach depths of 1 m, and finer sand forms loose drifts in the lee of prominent stones and in cracks in the solid substrate.

The western Jiddah (Fig. 4.2) is distinguished as an area of longitudinal features and shallow troughs near Haima, and blind drainage courses to the south. Oil industry operations have matched these drainage patterns with known subsurface structures, and they result from the progressive movement of deep salt layers and deposits. The surface of this zone is less rocky, with large expanses of coarse sand deposits which influence its vegetation.

The boundary between the central and north Jiddah is gradual, but is characterised by a decrease in all surface relief. The northern zone is a featureless plain of compacted sand, with localised spreads of limestone chips, rarely forming pavements. This zone and the distal fan both overlie the flat Miocene limestone, but the latter has a variable cover of pebbles from the northern pediments, resulting in vegetation typical of the pediments rather than the adjacent Jiddah without the pebble layer (4.5).

4.3   Hydrology

The Jiddat-al-Harasis is renowned in Oman for its waterless state through the absence of any natural water sources on the Miocene plateau and the poor ability of this surface to retain rainwater. This was the main determinant of the Harasis' former pastoralism (4.7). The quantity and quality of subsurface water sources still constrain development and the contemporary pattern of human use of the Jiddah, which is relevant to the oryx re-introduction.

Little rainwater infiltrates the various *hammada* surfaces, and modern surface runoff flows into the blind hollows, around which radial drainage patterns are evident. Rainwater rarely stands in the *haylahs* for more than 3 weeks and, although few have open sink-holes, many have signs of entry holes

Fig. 4.3. Traditional and modern water sources available to Harasis in the re-introduction area.

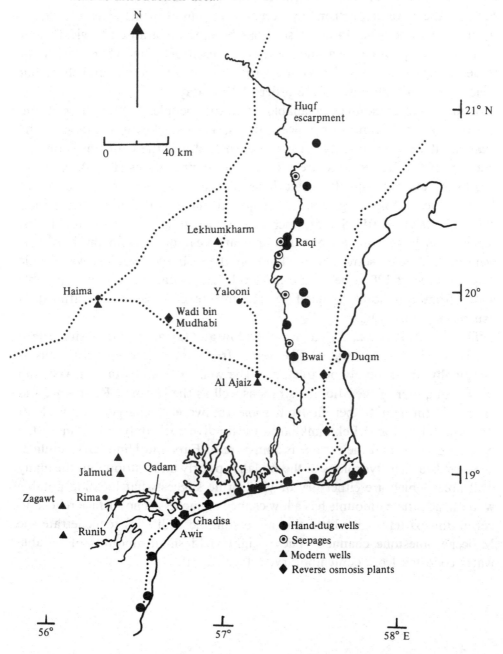

choked by wind-blown sand. These karstic sink-holes probably join a system of dissolution channels through the limestone.

The natural and artificial water sources of the Jiddah are shown in Fig. 4.3. In recent historical times, natural sources used by the Harasis were of two types. The first was a series of small hand-dug wells along the Sahel al Jazer coastline. The second was a series of springs, seepages and wells along the Huqf escarpment. As many originate only some 40–50 m below the Jiddah surface, their water is probably recent rain runoff which has percolated through the limestone. In doing so it has become mineralised to yield water that is barely potable even under extreme conditions. One characteristic of these springs is a marked seasonal variation in salt content and flow, but yields are invariably low, of the order of 1 m^3/day.

Modern water sources available to local people, rather than being specifically for oil camps or police stations, are also shown in Fig. 4.3. The main aquifer underlying at least the western Jiddah is the Umm er Radhuma (Parker, 1985). Its recharge area is the Dhofar mountains (Fig. 4.1), where this formation outcrops. Its water flows underground north-east at a rate of 12–13 m/year, following upon a major recharge in the pluvial period 8000–14000 years BP. Salinity increases with distance from the highlands, resulting in large volumes of water that are near-brine from the bore-hole wells of the Jiddah, some 400–500 km from the recharge area. The water table here lies at least 120 m below ground level. Large-scale provision of potable water from this aquifer, therefore, requires reverse osmosis treatment in expensive plants (Fig. 4.3).

The wells at Runib and Qadam are shallow and tap alluvial deposits in deep wadis to yield sweet water. The water from the other modern wells is marginally or seasonally potable only. The wells at Haima and Al Ajaiz tap the Fars aquifer of the Miocene strata as well as the Umm er Radhuma, and Haima is thought to benefit from some rainwater recharge. The wells at Zagawt, Jalmud and Lekhumkharm are for livestock only, even under the best conditions, and their water is a mixture of Fars and Umm er Radhuma.

The subsurface hydrology of the Jiddah is poorly understood, but the many drill holes which are either dry or yield brine suggest that locating potable water in adequate amounts has a low chance of success. The surface relief and vegetation patterns give little clue as to suitable drill sites to penetrate the Miocene limestone channels. These might yield small amounts of potable water collected from contemporary surface runoff.

4.4   Climate

The climate of the Jiddah displays the extremes associated with tropical deserts, but the proximity of the Arabian Sea exerts an ameliorating effect, with significant ecological consequences.

The monthly mean temperatures at Yalooni in the central Jiddah are shown in Fig. 4.4a. June is the hottest month, with an average shade maximum of 43 °C, and single daily extremes of 47–48 °C. July may be slightly cooler due to cirrus cloud intruding far inland from the southern highlands on the south-west monsoon. The coolest months are December to February, with monthly

Fig. 4.4. Average monthly maxima and minima at Yalooni, April 1980 to May 1986, for (a) shade temperature, and (b) relative humidity, with standard deviations.

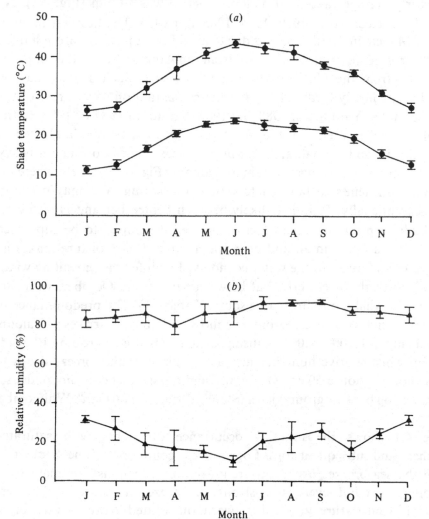

maxima of 26–27 °C, and an average minimum of 11.4 °C in January. The lowest temperature recorded has been 6.5 °C.

In addition to these monthly temperature changes, the temperature varies by 15–20 deg.C on every day of the year. Relative humidity often varies between less than 10 % and saturation in the course of 24 h, in winter or summer. There is less monthly variation in the maximum and minimum humidities (Fig. 4.4b). The high maximum humidities and low variance in July, August and September reflect the influence of the south-west monsoon, which also causes increases in the minimum relative humidity. There is no other clear seasonal pattern in minimum humidities. Variation around the averages for February, March and April is due to occasional periods of cold, dry northerly winds.

The yearly average rainfall at Yalooni 1980–6 was 44 mm (Fig. 4.5). No rain at all was recorded in 1980, 1984, 1985 and 1986. The heaviest monthly rainfall of 143 mm in April 1983 was due to a cold front passing through from the north. In general, rainstorms arrive from the Arabian Sea to the south or, more rarely, from the Gulf of Masirah to the east. Such storms may be intense, but are highly localised, so that there was rain on other areas of the Jiddah, but not at Yalooni, in 1980 and 1986, but not in 1984 or 1985. In the first half of 1986, areas in the south and south-west Jiddah received substantial rain, and a small area 25 km north-east of Yalooni had a heavy shower. From the occurrences of rain at Yalooni (Fig. 4.5) and elsewhere on the Jiddah, rain tends to fall in late winter and spring, and may occur in summer exceptionally. Rain is unlikely between September and November.

The Jiddah's perennial vegetation is too well developed to be supported solely by such a low rainfall and deep water table. Fog moisture makes an important contribution to the water economy. In winter the prevailing winds are either northerly or easterly, but between March and October, with rare exception, the wind comes from due south (Table 4.1). The predominance of this summer wind is due to a sea-breeze almost every day. It arrives at Yalooni 120 km inland at 16:00–18:00 h, causing air temperature to drop by 10 deg.C in 10 min, while relative humidity increases rapidly. If this breeze drops to below 8 knots (1 knot ≈ 50 cm/s) after midnight, the moist cool air condenses and forms a fog bank at ground level (Stanley Price, al-Harthy & Whitcombe, 1988).

Table 4.1 shows the monthly occurrence of nights with saturated atmosphere and subsequent fogs. Fogs are less common in June–September because the sea breezes are stronger, preventing the moist air mass from descending to ground level. These strong sea-breezes are caused by the very marked air temperature gradient between the heated Arabian interior of

Fig. 4.5. Monthly rainfall at Yalooni, 1980–6.

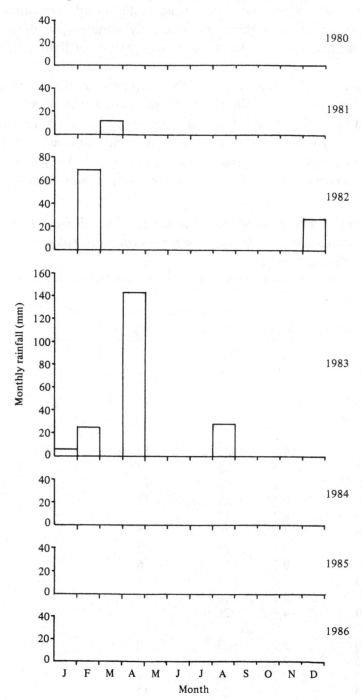

summer and the upwelling of cold seawater due to the south-west monsoon
(Stanley Price *et al.*, 1988). Some fogs occur in the winter months because of
convective heat loss under clear skies and relatively windless conditions
(Table 4.1). These winter fogs tend to occur during periods of light south-
easterly winds.

The relative amounts of fog moisture available to plants are measured at
Yalooni as the volume of water collected from galvanised iron sheets, of
0.25 m², suspended vertically to face due south and spaced between 0.9 and
4.2 m above ground. Over all months there is a significant increase in
moisture with height above ground (Stanley Price *et al.*, 1988). The volumes
collected are also influenced by the direction of prevailing wind (Table 4.2).

Table 4.1. *The direction and speed of prevailing wind, the number of nights
per month with saturated atmosphere and with fog moisture at Yalooni
between October 1983 and April 1985*

Month	Prevailing direction (degrees)	Speed (knots)	No. nights saturated	No. nights with fog moisture
1983				
October	180	6.0	26	11
November	120	4.4	27	3
December	330	6.0	9	2
1984				
January	30	5.2	20	1
February	90	6.5	22	12
March	180	7.4	23	7
April	180	7.4	18	11
May	180	9.5	29	12
June	210	8.7	17	1
July	180	12.0	30	3
August	180	10.2	25	2
September	180	10.7	29	6
October	180	6.2	21	15
November	120	5.2	26	14
December	360	4.6	26	9
1985				
January	120	5.2	23	10
February	330	7.9	12	4
March	180	7.5	22	11
April	180	8.9	20	12

1 knot ≈ 50 cm/s.

Moisture volumes are greater in spring and autumn months when the south wind prevails, and the height gradient is also more developed.

At Yalooni the southerly sea-breeze has travelled 120 km inland, while the south-easterly breeze has come only 70 km. The former is stronger and yields more fog moisture. This sea-breeze is detectable at least 75 km further north of Yalooni, but its frequency and that of fog so far from the sea are likely to be less. The influence of this breeze at Haima, 80 km west of Yalooni is less because of its greater distance inland due to the coast's shape (Fig. 4.2) and the influence of the nearby sand-seas on its climate. These effects, combined with the influence of the south-easterly breezes, with their lesser penetration across the Jiddah, result in a marked gradient in available fog moisture from south to north and from south-east to north-west. Yalooni is well situated to benefit from this combined influence.

4.5   Vegetation

The physiognomy, biomass and production pattern of the Jiddah's vegetation are determined by an annual average rainfall of less than 50 mm, with the possibility of several consecutive rainless years, and the amount of water available from the fogs. The distribution pattern of plants is largely determined by the water drainage pattern and the presence of adequate sand or fissures in the bedrock for plant establishment.

The density of trees on the central Jiddah often resembles an open parkland (Plate 2*b*). The most widespread tree is *Acacia tortilis*, which occurs as scattered individuals up to 5 m tall on ridges and the extensive, flat *hammada* surfaces. The tree shows several growth forms, and lichens, *Ramalina duriaei*, up to 2 cm long grow on any dead branch in the canopy. In any sandy depression and in all *haylahs*, *A. ehrenbergiana* is common. This species is multi-stemmed and straggly, rarely more than 3 m high. As each plant grows, a mound of wind-blown sand develops around its base, reaching maximum

Table 4.2. *The average amounts of fog moisture collected (ml/m²) per precipitation day at four heights above ground in months with prevailing wind from 180° and in all other months between October 1983 and April 1985*

Prevailing wind direction	No. of months	Height above ground			
		4.2 m	3.1 m	2.0 m	0.9 m
All directions	19	137	107	109	62
180°	10	176	137	135	49
Not 180°	9	99	77	83	73

heights of 1 m. In *haylahs* with deeper sand accumulations, single trees or stands of *Prosopis cineraria* grow to 10 m.

The most widespread and abundant grass is *Stipagrostis* sp., which occurs on all surfaces except in *haylah* sumps. Less frequent on ridges and in small drainage channels is *Dicanthium foveolatum*. The semi-prostrate *Ochthochloa compressa* occurs at up to 9 plants/m² (Lawton, 1978) in sandy areas liable to inundation. All three grasses can be annual or perennial, depending on conditions. Perennial grasses occur as tussocks in *haylahs* and on sand accumulations. The most common and important are *Lasiurus hirsutus* and *Cymbopogon schoenanthus*, while *Panicum turgidum* and *Pennisetum divisum* are more local.

Common perennial shrubs include *Rhazya stricta*, *Periploca aphylla*, *Heliotropium kotschyi*, *Crotalaria aegyptiaca*, *C. persica*, *Tephrosia apollinea*, *Limonium stocksii*, *Ochradenus harsusiticus*, *Pulicaria glutinosa*, *Zygophyllum* spp. and a variety of Chenopodiaceae such as *Cornulaca monocantha*, *Salsola* spp., *Halopeplis perfoliata* and *Suaeda* spp.

Smaller perennial forbs include *Heliotropium calcareum*, *Pulicaria undulata*, *P. jaubertii*, *Convolvulus* spp., *Dipterygium glaucum*, *Haplophyllum tuberculatum*, *Corchorus depressus*, *Kohautia retrorsa* and *Fagonia ovalifolia*.

After rain, annual grasses develop, including *Aristida adscensionis*, *A. mutabilis*, *Eragrostis ciliaris*, *Aeluropus lagopoides*, *Cenchrus ciliaris*, *Chrysopogon aucheri* and *Sporobolus* spp. Ephemeral forbs are also diverse, the most common being *Dipcadi erythreum*, *Arnebia hispidissima*, *Cleome brachycarpa*, *Euphorbia granulata*, *Monsonia heliotropoides*, *Indigofera philipsiae*, *Lotus garcinii*, *Lotononis platycarpa*, *Taverniera* spp., *Polygala erioptera*, *Kickxia* sp. nov., *Tribulus spp. and Zygophyllum simplex.*

Regional variations in species composition reflect the three geomorphological subdivisions of the Jiddah (Fig. 4.2) and the influence of the fog moisture gradients. In the south-eastern Jiddah, where this moisture is greatest, erect lichens grow to 3 cm on the ground. Plants such as *Lycium shawii*, *Cadaba farinosa*, *Cordia perrottetii*, *Pennisetum divisum* and *Panicum turgidum* grow no further west or north than the Yalooni area. Within 50 km of the south coast, the *Acacia tortilis* are wind-pruned, and the ground-level vegetation is less dense, because higher wind speeds here allow less fog to settle.

Within the central Jiddah there is a peripheral zone on its northern and western sides into which *A. tortilis* does not extend, probably because of moisture availability. *Prosopis cineraria* here is scarcer and the individuals shorter. This is transitional to the north Jiddah (Fig. 4.2), which is treeless apart from scattered *A. ehrenbergiana*. Large areas are dominated by the low, perennial shrubs *Zygophyllum qatarense*, *Ochradenus harsusiticus* and *Rhazya stricta*. The vegetation of the distal fan is the same as that of the lower northern pediment fans, with a rather denser cover of low *A. ehrenbergiana* with small sand-mounds than on the north Jiddah, *Heliotropium kotschyi* and dense annual *Stipagrostis* sp. after rain.

The western Jiddah also has no *A. tortilis*, but *P. cineraria* grows in groves in the low-lying sumps with deeper sand. *A. ehrenbergiana* is abundant. Species composition changes rapidly westwards with development of the gypsum-rich surface. This supports no trees or bushes, but characteristic forbs such as *Fagonia ovalifolia*, *Cornulaca* sp. and *Salsola* spp. over very large areas.

The gradient of increasing amounts of fog moisture up to at least 4.2 m above ground level probably causes the abundance of trees on the Jidda' despite the low rainfall. The roots of *P. cineraria* can tap water from as deep as 36 m (Leakey & Last, 1979), but the water table of the Jiddah is probably out of reach. The common sight of a young *Lycium shawii* or *A. tortilis* establishing itself within 2 m of the base of a fallen dead tree suggests competition for sites where the underlying limestone has already been cracked by roots. Moreover, the *A. tortilis* of the Jiddah are easily blown over by wind, revealing a dense network of shallow surface roots but no deep tap-root.

Under experimental conditions, both *A. tortilis* and *P. cineraria* absorb condensed water through stomata (P. Gates, personal communication 1985). As this occurs mostly in the dark at low temperatures, the trees are adapted to using fog moisture. The *Acacia* spp. may also absorb through their roots water which has condensed and dripped from their canopies.

The trees of the Jiddah show some green leaf at all times of the year. In the absence of rain, there are bursts of new leaf in March–April and in September. As air temperatures are then moderate and fogs most common, most perennial forbs, shrubs and grasses respond to this moisture. Growth, rather than germination, is stimulated, and grasses flower. The main ecological consequence of the fog is that the Jiddah has a background level of primary production twice a year. This predictable regrowth maintains grazing quality long after the main response to rain, in addition to stimulating light growth in areas without rain for many years. This regrowth greatly benefits the area's herbivores, including the oryx.

4.6   Wildlife

4.6.1   *The present wildlife community*

The Jiddah's well-developed vegetation meets the food and habitat requirements of a diverse fauna. Full taxonomic names are given in Appendix 1. Although the oryx is the largest herbivore, the most numerous is the Arabian gazelle. Found all over the Jiddah but with most in the central and southern zones and down into the Huqf and coastal plain, the species has increased much with full and effective protection since 1976. The population is the largest in Arabia, and is estimated to number 5000, with upward fluctuations after good rain. The goitered gazelle lives in the sand-seas and peripheral plains. Its numbers are unknown and its habits barely studied, but its habitat is more extreme and overlaps little with that of the smaller Arabian gazelle. The Huqf escarpment and hills to its east support a small but viable population of Nubian ibex. Hares are found throughout the Jiddah, but their population level fluctuates widely with rainfall. Hedgehogs are widespread but rarely seen.

The Arabian wolf used to be a serious predator of livestock before cars were used to pursue them. The species now lives in the shelter of the Huqf and southern escarpments and wadis. It ranges across the Jiddah only in winter when the climate allows surface dwelling and long distances to be covered to avoid man. Predation by wolves still deters the Harasis from settling in some wadis leading to the coastal plain. The striped hyaena is not known from the Jiddah but predates goats and scavenges fish seasonally along the coast. Resident carnivores on the Jiddah include the caracal, which predates gazelles, the wild cat, and the honey badger, which scavenges carcasses but primarily excavates burrowing lizards (see below). Rüppell's sand fox is native, but the congeneric red fox is spreading with developing human settlements and displacing the smaller desert species around them.

The monitor lizard is active in spring and after rain. The other large lizards comprise two herbivorous species of the genus *Uromastix*. The smaller, the spiny-tailed lizard, excavates holes about 25 cm deep through stone chip substrates. The larger, *dhub* in Arabic, occurs only where there is enough superficial sand for its burrow, as in the west Jiddah, and on the escarpment where rock crevices provide secure shelter. The commonest snakes are the venomous horned and carpet vipers, the hooded malpolon, the sand snake and cat snake. Smaller reptiles include a variety of skinks, agamids and geckos.

The commonest rodents are the gerbil and Sundevall's jird. The lesser jerboa and spiny mouse are much less common. The population levels of the first two species increase explosively after rain. These rodents and the burrowing reptiles all live in the sand-mounds around *A. ehrenbergiana*.

Bird species recorded on the Jiddah over 6 years total 168 species (K. O. Stanley Price, personal observation). Twenty-two species are breeding residents, and include the golden eagle, houbara bustard, spotted thick-knee, coronetted and chestnut-bellied sandgrouse, black-crowned finch-lark, hoopoe lark and crested lark, great grey shrike, little owl and brown-necked raven. A further 15 species are known as winter visitors and 104 species seen as passage migrants in spring and autumn.

4.6.2 *Aboriginal use of the Jiddah by oryx*

The accurately located sightings of oryx from the diminishing population in the 1960s (Fig. 2.3) show that oryx were found in all three subdivisions of the Jiddah, but the northern zone was perhaps less favoured (Fig. 4.2). The older Harasis confirm that certain larger *haylahs* and drainage lines of the central zone almost always supported oryx. The greatest numbers were universally held to have been on the open plains and broad wadis of the west Jiddah, and it was from here that most were captured in the 1960s (2.6).

The oryx were said to move into the sand-dunes after rain and in winter, exploiting the superior grazing there (Thesiger, 1959; Loyd, 1965), but had to move southwards again when shading became necessary. This move brought them into the area favoured by sea-breezes (4.4). The oryx tended to avoid humans, using the shade trees of the central and eastern Jiddah during the summer, which the *bedu* vacated for denser shade close to the permanent water sources of the Huqf (Fig. 4.3). When the *bedu* returned to the Jiddah in winter, the oryx retreated to the western Jiddah, where the grazing was said to be particularly rich and human disturbance was less. This area is now the part of the Jiddah most subject to oil industry activity (4.8.1). The implications of this for the re-introduction are discussed later (9.3.1).

Written and oral records of oryx locations indicate that the original population was largely water independent, although the re-introduced animals have shown that individuals could walk 30–40 km for a drink overnight and be back at the same place by morning (6.9). Although there are no definite records of their using Huqf seepages, footprints at the brackish creeks of Awir and Ghadisa (Fig. 4.3) were recorded. Oryx from the west Jiddah used camel tracks to descend the Wadi Runib to the coastal plain. In 1986 oryx faeces from the original population were found under a cliff overhang in this wadi.

4.7   The Harasis people and their environment before 1970

The origins of the Harasis tribe are not known. However, although not related to the Mahra tribe of southern Oman and eastern People's Democratic Republic of Yemen, their traditional languages belong to the same pre-Islamic, south Arabian group (Johnstone, 1977), and the tribe must have moved from south-west Arabia. There is no evidence of the reasons for their occupancy of the Jiddat-al-Harasis nor for the date of their arrival, although this is likely to have been several centuries ago.

Before the advent of modern development in Oman in 1970, the way of life of the Harasis (singular: Harsusi) was determined by the remoteness of their lands from any permanent settlement, and their natural resources. As nomadic pastoralists, they followed rain and grazing over a vast area of central Oman, between the eastern wadis of the southern pediment and the wadis draining the northern mountains (Fig. 4.1). Despite the perennial vegetation of the Jiddat-al-Harasis, the lack of water deterred many Harasis from ever living on it; they preferred to use the lower reaches of the Wadi Halfayn. This wadi is the major water course draining the inland face of the Hajar mountains and runs into the sea at the Hikmann flats (Fig. 4.1). In the lower, northern fans, the wadi is fringed with *Prosopis* woodlands and has sweet water near the ground surface. Some families, particularly those with elderly members, lived almost permanently near the Huqf's two largest water sources, Bwai and Raqi (Fig. 4.3), while many others moved up on to the Jiddah after rain there or in winter when water requirements were lower. A few families were renowned for occupying the Jiddah year-round, based on the larger *haylahs* with their constant *Acacia* browse. Their goats were managed for water independence, even through summer, while the people subsisted on very small amounts of condensed fog moisture in addition to milk.

Except under more relaxed conditions immediately following rain, a Harasis family group on the Jiddah spent only 1–2 days in each *haylah* while the goats harvested the small amounts of green growth, before moving on (Harasis, personal communications 1986). This mobility demanded a minimum of material possessions, and a family had a few blankets, one cooking pot and a coffee pot, and a woven leather milk bowl. Unlike most *bedu* tribes, the Harasis had no tents for shade, using instead the umbrella-shaped *Acacia tortilis*, and this reduced the possessions that had to be transported.

The family's livestock survived for most of the year without free water to drink. Their goats were a short-haired pure white breed, whose low physiological water requirements were reduced in summer by night grazing and walking. An average family had 60–100 goats, but some had up to 300 head. About six camels was normal for transport. Sheep were much scarcer, reflecting the poor opportunity for weaving in a society that was constantly on the move.

Human water requirements were very low, and milk was the main source of fluid. However, men rode camels for up to 5 days to fill goat-skins with water at Bwai or Raqi before returning the same distance, or travelled to the coastal wells if they lived in the west Jiddah (Fig. 4.3). The water carried by one camel lasted a family no more than 5 days. In addition, drops of condensed fog moisture were harvested from the canopies of *A. tortilis* trees, by spreading a blanket or length of special material under the tree and tapping the branches. The water was then wrung from the material for drinking.

With the exception of a few commodities, the Harasis' dietary requirements were met by milk and meat. The livestock were managed actively for these products. The total goat flock was divided into three groups of males only, females in milk with their kids without any breeding males, and dry females running with males. Animals were moved between groups as necessary, and the breeding rate was adjusted to local grazing conditions and to ensure that a female lactated for seven to eight months. The separation of these groups required active management by the bedu, and the women walked all day with their flocks to prevent straying and predation, and to ensure that the goats returned home in the evening.

The Harasis actively exploited their environment for better stock management. The lower branches of suitably shaped *A. tortilis* trees were lopped off or half-cut through and bent down to make a tent-shaped shelter. Inside, goats were better protected from strong winds, hot or cold, and direct sunlight. Branches were cut to form thorn enclosures as protection from cold winter winds and to prevent night-straying. Early on winter mornings the canopies of *A. tortilis* were beaten and the new, fallen leaf collected from blankets and fed to camels and goats as a supplement. In season, pods of this species were collected for the same purpose and traded between Harasis.

Camel management was also highly developed. Milk production may have been subordinated to ensuring enough pack camels for the family. Thus, some females were not allowed to breed, and not all households invariably had camel's milk. In addition to the camels that travelled with a family, each had potentially far more camels at free-range. Each camel was branded with its tribal and owner's mark, enabling the latter to hear from other tribesmen when and where it had calved. Under these circumstances the female and calf would be 're-domesticated' as an extra source of milk.

Although goats, particularly surplus male kids, were the main source of meat, wildlife was the preferred meat source in many households. Men hunted singly with a camel for as many days as they could carry their own supplies. The main prey on these forays were ibex, gazelles or oryx, but small animals such as hedgehogs, hares and the spiny-tailed lizard were also taken. As water was too short for boiling meat, it was grilled on flat stones that had been heated in a fire made from the abundant dead wood of the Jiddah.

Trade and barter were needed by the Harasis. Groups of men accumulated enough goats to be walked to the markets of northern Oman, taking a month to reach them by travelling at night, as necessary, and preferably after rain along the route. Some journeyed to the markets with camel loads of salt hacked from mines between the Jiddah and the villages. The main items bought were rice, coffee beans and dates, and material for new clothes. More locally, the Harasis bartered or sold in-milk goats to the Janabah fishermen with no flocks of their own on the east and south coasts. They also took to the coast ropes plaited from the dwarf palm, *Nannorhops ritchiana*, found in the wadis of the Huqf, to be used on fishing nets, and loads of firewood. The Harasis returned with dried fish, mostly shark, for their consumption, and, in season, sacks of beach-dried sardines as a supplement for their livestock. The coastal tribe, the Janabah, were also able to supply coffee, dates and other commodities which had been shipped from the Muscat coast.

The number of Harasis before 1970 can only be guessed as around 500. While child mortality was undoubtedly higher than now, their numbers may have been limited by a low birthrate, for families with more than two to three children were said to be rare. It is obvious, however, that population density was very low, as the Harasis on the Jiddah practised a form of pastoralism which was highly specialised even by Arabian standards. Because of their mobility, lack of material possessions and the distribution of water, the Harasis' impact on the Jiddah was very light. Under these conditions, they were dependent on their livestock for most of their requirements. In such a marginal environment, this demanded highly competent stock management.

4.8   Development after 1970

4.8.1   *Infrastructure*

The accession of HM Sultan Qaboos bin Said in July 1970 started a development drive throughout Oman, fuelled by accumulated oil revenues. Current revenues also increased fast because of greater extraction and the world-wide increase in the price of oil in 1973. Despite its remoteness and distance from both Muscat and Salalah (Fig. 4.1), the Jiddat-al-Harasis enjoyed its share of development, and little of the infrastructure present in 1986 had existed 16 years before.

Fig. 4.6. Infrastructure of the re-introduction area in central Oman.

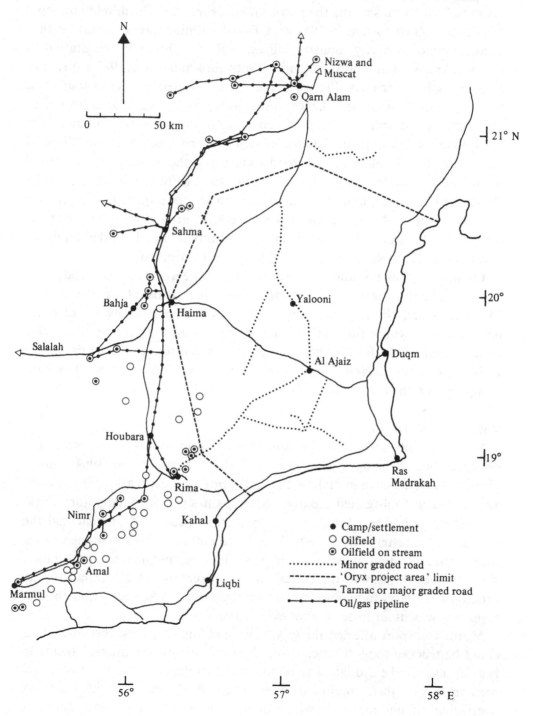

N

0   50 km

Nizwa and
Muscat

⊙ Qarn Alam

21° N

Sahma

Bahja

Haima

Yalooni

20°

Salalah

Al Ajaiz

Duqm

Houbara

19°

Ras
Madrakah

Rima

Nimr

Kahal

● Camp/settlement
○ Oilfield
⊙ Oilfield on stream
········ Minor graded road
------ 'Oryx project area' limit
——— Tarmac or major graded road
●–●–● Oil/gas pipeline

Amal

Liqbi

Marmul

56°   57°   58° E

The Sultanate's main arterial tarmac road from Muscat and Nizwa in the north to Salalah runs across the north Jiddah (Fig. 4.6). The developing town of Haima was completed in 1981 as a Tribal Administrative Centre, with a local governor, priority ministry offices, a Royal Oman Police Station, a petrol filling station, shop, restaurant, mosque, and since 1982 a primary boarding school for boys. From Haima, a main graded road runs south-east to the well of Al Ajaiz and thence to the fishing village of Duqm. This road then runs northwards parallel to the coast, returning to northern Oman. On the Jiddah, 20 km short of Duqm a graded road runs south to the village of Ras Madrakah, while a branch road runs along the Sahel al Jazer coastal plain as far as Liqbi village. A graded road runs from the Sahel al Jazer to Rima oil camp. An unmotorable graded road from the 1950s runs past Yalooni from Al Ajaiz, but there is no modern, graded road access to the camp. Over the Jiddah there are many blind, unmaintained, graded roads for one-time access to specific locations by oil exploration rigs.

The major infrastructural developments are to the west in the oil production areas. Currently there are important producing fields along the axis of Marmul, Amal, Nimr, Rima, Bahja and Sahma. A dense network of roads has developed within these fields, while a main road follows the oil pipeline from Marmul to the fields of northern Oman. In addition to residential camps with full services at these fields, there are police stations at Rima, Houbara pumping station, Haima and Sahma.

4.8.2   *The Harasis*

Before 1970 the Harasis' cash requirements were low, and opportunities for employment few and short term. From 1964, survey companies were active in making the first topographical maps of the Jiddah, and seismic and oil-related exploration companies provided temporary work. In the 1970s the bedu benefited swiftly from modern government and the buoyant oil industry. Their livestock, hitherto almost worthless in cash terms, rose in value, and oil industry companies offered employment opportunities. The new wealth allowed the Harasis living in remote central Oman to buy cars. By 1975 every family was running at least one truck, and within another 5 years it was usual to see two at every camp.

Motor transport affected the way of life profoundly. First, because water could be trucked long distances from the few sources, the limiting factor to heavier use of the Jiddah was removed. Most Harasis would rather live permanently in their traditional home, the Jiddah, in the absence of any overriding commitments elsewhere or attraction to other areas. This, for example, occurred partially in 1982–3, and in 1987 when the rainfall pattern caused almost total vacation of the Jiddah in favour of Wadi Halfayn, 200–250 km north-east of Yalooni (Fig. 4.1) Although this has always happened probably once every few years, the total impact on the Jiddah from people and livestock increased through the 1970s. Second, because camels were no longer needed for transport, they could be managed for milk production exclusively. Despite the slaughter of many male calves at birth or at weaning, their numbers may have increased. Every family now has at least two females in milk at any time, and their milk is preferred to that of goats. As the Harasis gradually transported drums of water to their camps by vehicle, soon aided by a Government tanker service, there was drinking water every day for the goats and camels. Under these conditions in the northern Jiddah, the white goats were seen to be less productive than the long-haired black goat of the better-watered Wahiba areas. As the white goats were also highly susceptible to pneumonia in winter, the black goats had replaced the white in most Harasis flocks on the northern Jiddah by 1980.

The availability of cash in every home promoted great improvements in welfare. Foods were bought in the villages: rice, onions, dates and coffee became daily staples, and ghee, oil, tea, tinned milk, sugar and tinned meats were also used. While the *bedu* slaughtered fewer goats, the new foods and continuing, abundant milk produced a high quality and balanced diet (Chatty, 1984). Through the 1970s the quality of housing started to improve. Tents were acquired but after a few moves they were used more as tarpaulins to extend the shade of a tree or for wind-protection. In 1984 the first sectional frame dwellings appeared (Plate 4*a*). Using tarpaulins tied down, the degree of shelter and shade could be adjusted infinitely, and the efficient exclusion of goats from inside improved comfort and hygiene. The disadvantage of these dwellings lay in more labour to dismantle and pack them across a truck bed. Thus, this development in housing and the acquisition of household belongings discourages frequent moves, and a camp-site may now be used for several months.

The numbers of goats per family increased because of bringing water to the camp and because of less offtake due to the modern diet. When the natural grazing became limiting, the Harasis response was to buy pelleted concentrate supplement. As a better-balanced feed than the traditional sardines, and with year-round availability, the demand for pellets increased sharply in the 1980s. In the rainless periods from 1984, concentrates for livestock became most families' major monthly item of expenditure (8.2.2).

While goat numbers increased for these reasons, there has been no accompanying increase in the traditional means of adjusting numbers to carrying capacity, namely through sale. Sales are currently a last resort solution to raise cash, and so only kids or adults in poor condition are marketed. The Harasis are disadvantaged by the 300–400 km distance to the markets of northern Oman and by the fact that in each their stock has to compete with higher-quality goats brought in by the *bedu* of the nearby mountains. This re-inforces the feeling that earned salaries have replaced the need to sell stock regularly to obtain cash (Chatty, 1984).

These recent improvements in living standards and diet have been accompanied by changes in livestock management. Because human survival no longer depends on optimising milk and meat production, all the goats of most families run together. The resulting near year-round dropping of kids is divorced from an optimally timed, synchronous breeding-season following rain and fresh grazing. In addition, because the goats are now conditioned to return to the campsite for water and supplements, no one walks with them while grazing. Thus, the goats lose the benefits associated with the presence of a herder, and the problem of small groups straying and having to be tracked up by vehicle and retrieved is exacerbated. The productivity of individual animals for meat or milk is less under these management conditions.

In conclusion, it is clear that, except after rare periods of rainfall, when huge areas may remain ungrazed, the increased number of goats no longer meets the Harasis' needs for milk and meat. Rather, because of the high inputs of extra feed needed to maintain the greater numbers, the livestock are dependent on the *bedu*. This relationship, reversed from the pre-1970 era, is the single factor which ensures that a large proportion of Harasis men must continue to find work in the cash sector of modern Oman.

4.8.3   *Present population and employment*

The low densities and mobility of the Harasis make a formal census difficult. A detailed census in 1981–2 of the family units living on the Jiddah alone showed an average nine members to a household. With the families living at the time in the wadis to the north-east, the tribe probably totalled 1800 people (Chatty, 1984). Its numbers are increasing for the census also showed 19.2% to be in the 0–4 year-old class, with a further 14.6% aged between five and nine years. This population increase is further reason for the declining role of livestock in supporting families.

Table 4.3 summarises the sources of employment in 1986. The oryx project, with 26 Harasis staff, is second only to the national oil company, although the nature of the work in the two is regarded very differently (8.7). As males between the ages of 15 and 54 account for 16.7% of the present population (Chatty, 1984), the 142 employed represent 47% of the potentially employable adult males. This is consistent with the estimate that two out of three of all

household heads are regularly employed (Chatty, 1984). The remainder
constitute those engaged fulltime in the tribe's pastoralism and are responsible
for the families of male relatives who may be working far away.

4.8.4 *Harasis' attitudes to their environment and conservation*

The Harasis of the pre-motor age were highly competent pastoralists, with a
profound understanding of how to exploit the sparsely endowed and unevenly
distributed resources of the Jiddah. The peculiar water economy forced on
them for most of the time ensured that they were specialists even among the
pastoral *bedu* tribes of Arabia. Many of their methods of resource use show
their natural conservation policies:

   (1) When gazelles, ibex or oryx were hunted, the group size determined
       whether only one or more could be taken. This was done specifically
       to ensure the future of these resources. The dead animal was then
       used fully, to the extent that the rumen juices of an oryx would be

Table 4.3. *Sources of employment for the Harasis in 1986*

Employer	Posts	No. employed
1. Ministry of Interior	Shaikhs, guards, patrols	17
2. All other ministries	Guards, drivers, guides, water tanker operators	21
3. Royal Oman Police	Civil Guards, guide, camel patrol	16
4. Petroleum Development Oman	Staff of Depts of Production, Fire and Safety, Maintenance, Engineering	30
5. Arabian oryx project	Rangers, drivers, workshop, kitchen	26
6. Selfemployed[a]	Traders, contractors garage-owner	6
7. Companies	Drivers, seismic crews, temporary guards	21
8. Various in United Arab Emirates		5
Total		142

[a] In addition to the six selfemployed, a further four men included in other categories
have private business interests.

drained and drunk. With such a large prey, the hunter would camp at the site until all the meat had been dried in strips and could then be transported by camel to the family camp.

(2) No living tree is cut to provide firewood. Dead branches only are broken off by hand from within trees. If branches of *A. tortilis* have to be cut to form an enclosure, only the lower branches are taken, so that the tree's shade is not reduced. *Prosopis cineraria* is lopped to provide fodder for sick livestock only, and the Harasis know that a tree must be lopped every 1–2 years to maintain vigorous growth. New shoots of this species can grow 6.5–23 cm within 2 months of lopping in eastern Oman (Brown, 1986), and the Harasis say the lopping must be done with an axe rather than by breaking of the branch.

(3) If one family has reserved a campsite by hanging a few belongings in a tree until they can move in, anyone arriving subsequently will not camp in the immediate vicinity, but will move off to reduce competition and the impact of livestock. Similarly, anyone finding ripe fruit on a wild date palm at a spring must only take one bunch, leaving the rest for other travellers.

(4) Under some conditions of rainfall when it has been neither very widespread nor so local that the new growth must be utilised, Harasis men will define and declare an area a grazing reserve. All settlement and grazing by goats are prohibited within it. The area is for the benefit of camels and wildlife and is done in the interests of rangeland management. Its reserve status lasts until the camels disperse because grazing conditions are the same inside and out.

(5) It is of great significance to the oryx re-introduction that the Harasis of old knew that no other tribe's lands supported oryx like the Jiddah. Their pride in this and the possessiveness of the paramount shaikh were well known (Loyd, 1965), and the loss of the oryx was much regretted soon after their extermination.

Despite the many developments in the Harasis' household economy and pastoralism since 1970, these attitudes still persist and are apparent in everyday life of the *bedu*. On the other hand, their new way of life has brought its own environmental problems, mostly deriving from uncontrolled vehicle tracking through *haylahs* and other prime grazing areas. With the effects of households using the same site for months instead of days and the greater occupancy of the Jiddah by the Harasis, the vegetative cover has decreased. It is a cause for concern to many tribesmen.

4.9   Conclusions

This chapter has shown that the Jiddat-al-Harasis has well-developed vegetation and a diverse community of wildlife species, despite very low rainfall. The contribution of fog moisture results in a distinctive and unusual desert ecosystem, which is worth preserving for its productivity. The area's terrain, vegetation physiognomy and climate influence in many ways both progress towards successful re-introduction of the oryx and the management methods used.

Human occupancy of the area until the recent past was limited by the availability of water. Although limited water development has initiated major ecological changes, the area's hydrology suggests further water sources may be developed only at great cost. Although the Jiddah's water sources remain few and unevenly distributed, the present dependence of Harasis livestock on supplementary feeds suggests that over the last 15 years the amount of grazing, rather than water availability, has come to limit their animal production. As these changes took place in the precise period between the extinction of the original oryx and their return, even a superficial treatment of Harasis ecology shows that the oryx habitat must have changed during their 10 year absence between 1972 and 1982. Although the feasibility study discounted severe habitat change over this period (Jungius, 1978b), the effects described here raise, at least, the possibility that the new oryx population might not exploit its environment in the same way as its predecessor.

The Harasis *bedu* knew how to use their environment in a sustainable fashion. They are still concerned for its wildlife, confirming them as ideal guardians of the re-establishing oryx in their tribal lands. Because the Jiddat-al-Harasis has no formal conservation status, their knowledge and attitudes may have been especially critical to the project's viability. Subsequent chapters will provide ample illustration of the Harasis' vital contribution to the project's success.

5

Oryx management

5.1   Introduction and aims

Chapter 1 contained several cautionary examples of the reaction of translocated animals upon release into a new area. The management of the oryx in Oman from their arrival in the country was carefully planned to avoid such mishaps. This chapter describes the management aims and methods, and the process of monitoring the oryx in the three main headings of pre-release, the process of release, and the post-release era while the animals adapted to the unenclosed desert.

The moment of release is the most critical period in translocation or re-introduction operations. Although it may represent a symbolic end-point for the manager and general public, it only marks the beginning of the animals' most testing time. The process of adaptation, leading to genuinely successful establishment, requires years and, perhaps, several generations following release. So, management of the oryx was aimed at preventing explosive scattering of herd members on release. Once scattered, it would probably be impossible to trace all of them, and a single animal lost in a new habitat is likely to perish (1.3.2).

5.2   Pre-release management
5.2.1   *Strategies*

To ensure successful release and survival of oryx, pre-release management was designed around five strategies.

(1)   The imported animals had to acclimatise to the more rigorous and extreme climate of Oman. Summer maximum temperatures at Yalooni are much higher than in San Diego, and also greater than at Phoenix (Fig. 5.1). Winter maxima are similar at San Diego and

Phoenix but lower than at Yalooni. Winter minima are higher at Yalooni than both US sites. However, summer minimum temperatures are highest at Phoenix, followed by Yalooni and then San Diego. Rainfall at San Diego in 1985 was six times that of Yalooni's average over 7 years.

(2) The oryx had to habituate to the sights, sounds and smells of their new environment and its people, other animals and vehicles. The aim of this habituation was to discourage scattering on release and, then, to allow close approach to the oryx for accurate monitoring.

(3) The gregariousness of all oryx species indicated that the basic unit for release should be the herd rather than the individual. The optimum group for release was considered to be the 'integrated' herd, which would total not less than 10, contain both sexes in approximately equal numbers, contain a range of ages of each sex,

Fig. 5.1. Average monthly (*a*) maximum (shade) and (*b*) minimum temperatures, and (*c*) rainfall at San Diego Wild Animal Park (SDWAP) (●), Phoenix Zoo (▲), Yalooni (♦).

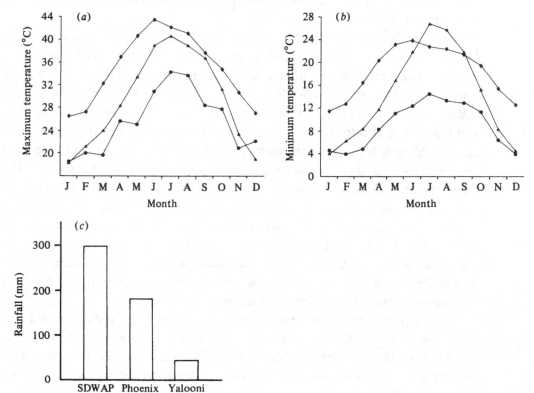

and its members would have been together long enough for a stable hierarchy to have developed amongst them (2.2).

(4) The oryx had to form an attachment to the release area so that it had a pivotal position in their future home-range in the wild. Conversely, if the oryx were determined to escape the immediate release area, this would negate the careful selection of the re-introduction area as the best oryx habitat available to the project. This conditioning of the oryx was achieved by provision of supplementary feed and water in the release area only (5.4.7).

(5) As few changes as were within the control of management were to accompany the transition from captivity to life in the open desert. With minimal changes, the chances of running and scattering would be reduced. This consideration influenced the optimum ecological conditions for release and the actual process (5.3.1).

5.2.2   *Preparations for importation into Oman*

Preparations for oryx travel from the USA to Oman fell into three categories: (1) permission to export from the USA, (2) the action needed to meet Oman's veterinary health import requirements, and (3) the logistics of the transport itself.

Each individually identified oryx (5.2.9) required a US Department of the Interior permit for export, backed up by a Federal Fish and Wildlife permit to meet the requirements of the Convention on International Trade in Endangered Species of Wild Fauna and Flora (CITES) because of the endangered status of the Arabian oryx. The California Department of Fish and Game had also to issue an export permit which specified the numbers of animals of each sex in the consignment, and this same department inspected the animals and their accommodation at the airport of departure. The Department of Commerce had to issue a Shipper's Export Declaration, which included precise details of the flights and the documentation. In view of the oryx's conservation status, the Fauna Preservation Society as a Trustee of the World Herd had to confirm to the Secretary of the Interior that it approved the export of oryx to Oman. The government of Oman had to declare that it was willing to participate in a co-operative breeding programme of the oryx and would maintain a studbook of animals in Oman.

Oman's veterinary health requirements are shown in Appendix 2. Waivers were given by Oman with respect to foot-and-mouth disease which is unknown in the US, and anthrax for which vaccination is only permitted following an outbreak. As blue-tongue is endemic in the US, no assurance could be given that the area was free of this disease. Moreover, during the testing of the first group of 12 oryx destined for Oman, 7 animals were positive on the blue-tongue serology test, although none had shown any clinical symptoms of the disease. Subsequently, after all five oryx of the third group for Oman were found to be positive on the serology test but negative on a virus isolation test, the latter was accepted as adequate for import into Oman.

Preparations for transporting the oryx started in the USA with their immobilisation and removal from their herds to start a 45-day quarantine (Appendix 2) and undergo tests. Each oryx travelled in its own crate of dimensions that allowed it to stand and lie, but not to turn. The height of the crates ensured that there was little chance of damage to horn tips. Hay was provided both to feed and to lie on. The animals were loaded through a crush without any sedation. The crates were trucked to Los Angeles airport and then by commercial cargo airliner to Oman, usually via Europe. On arrival at Muscat International Airport, the crates were transferred to military transport aircraft for immediate onward shipment to Yalooni. The time spent crated varied between 40 and 95 h for various groups, depending on aircraft scheduling and connections. Even after 95 h, the oryx were not adversely affected, and did not appear to be water-stressed.

5.2.3   *Oryx received in Oman*

The original 12 oryx selected for Oman comprised six of each sex, with the males aged, at 1 January 1980, 37, 21, 17, 16, 8 and 7 months old, and the females 23, 23, 18, 17, 11 and 10 months. Following substitutions due to positive blue-tongue tests and the consequent delay in preparing further groups, the actual animals sent to Oman had a wider age distribution (Table 5.1).

A total of 18 animals arrived in Oman, of which 17 came from the USA, and the other from Jordan (Khalifa). Of the 17, two (Museba and Rahaima) came from Gladys Porter Zoo, Brownesville, Texas, while Kadil came from San Diego Zoo. The remaining 14 came from San Diego Wild Animal Park. With the exception of group 3, all the animals arrived at ages between 9.5 and 27.5 months. Farida, arriving at 43.5 months, was the only oryx to have bred in the USA, having had two calves. Both Salama and Rahaima travelled pregnant, calving 2.2 and 5.7 months, respectively, after arrival in Oman. Full information on each animal is given in Appendix 3.

The commitment from the World Herd of six males and six females was met with group 3, but this group contained an extra two males. As this led to a preponderance of males in Oman, SDWAP agreed to provide the three young females of group 4, following requests to this effect from WWF International and IUCN.

Only when the first oryx had arrived at Yalooni were the Harasis satisfied that the animals being returned were, indeed, the same as those they had known before. There were, however, differences between the animals. The Brownesville pair, which were genetically from Phoenix stock (Table 3.6), resembled one another but differed in conformation from the SDWAP animals. The former were shorter in the leg, had relatively sloping and rounded hindquarters, and long, pointed ears. As they were related, their only calf, Alaga, was highly inbred (7.5.2), and retained some of these characteristics to a lesser degree. Some of the differences in appearance were, therefore, probably genetic, but the conditions of their zoo upbringing must also have been responsible. Their conformation has not persisted in Oman through the relative failure of this line (7.4.1).

Table 5.1. *Details of oryx imported into Oman*

Group	Date of arrival (day/month/year)	Name	Sex	Age at arrival (months)	Months spent in Pens	Enclosure	Herd in Oman
1	6/3/80	Jadib	m	22.0	1.9	20.9	1
		Salama	f	20.6	1.9	20.9	1
		Talama	f	11.6	1.9	20.9	1
		Hamid	m	10.1	1.9	20.9	1
		Malak	m	9.5	Died	—	—
2	8/12/80	Nafis	m	27.5	3.4	10.4	1
		Museba	m	21.2	2.5	11.3	1
		Rahaima	f	20.6	2.5	11.3	1
		Sajba	f	19.6	3.4	Died	—
		Hadya	f	15.4	2 days	13.7	1
3	8/9/81	Kadil	m	113.1	Not released	—	—
		Lubtar	m	49.5	7.1	23.7	2
		Farida	f	43.5	7.1	23.7	2
		Mustafan	m	37.3	20.4	10.4	2
4	15/7/83	Zeena	f	17.9	2.7	5.9	2
		Mustarayha	f	17.1	Not released	—	—
		Hababa	f	14.7	2.7	5.9	2
5	25/1/84	Khalifa	m	11.9	0.5	1.8	2

m, male; f, female.

5.2.4 *Oryx facilities at Yalooni*

The facilities built for the oryx consisted of a set of interconnecting pens, and an adjacent enclosure with natural vegetation (Fig. 5.2). The pens were not designed for permanent occupation by oryx, but for as short a period as possible before the integration process in the enclosure. The external fence of the pens was continuous with the enclosure's perimeter fence, and was made of 2 m chainlink wire, with three horizontal, high tension cables woven through the mesh. This was supported by 3 in. galvanised iron pipes at 3.3 m intervals. Each upright was sunk into the base rock and cemented. At its upper end, there was an out-turned cranked section of 0.6 m, carrying two strands of barbed wire. The bottom of the chainlink was just below the original ground level, but overlapping with it to a height of 0.3 m was a strip of plastic-coated chainlink, which lay in a trench 0.3 m deep, before turning out horizontally for 0.6 m. This fence was, therefore, likely to deter any burrowing animal and any unauthorised person from climbing in, and no oryx can leap 2 m. Access into the enclosure was through two gates adjacent to and inside the pens, and one in the diagonal opposite corner, through which animals were released (5.3.2 and Fig. 5.3). Once the oryx were seen to be secure in the enclosure, stepladders were put over the fence on the east and west sides for easier access.

The pens lay on the highest ground of the enclosure, at its north-west corner (Fig. 5.3). All internal divisions were made of the same chainlink and supports, but without underground extensions or barbed wire on top. The main accommodation was two 20 m × 20 m pens, each with free-standing

Fig. 5.2. Layout of oryx pens at Yalooni.

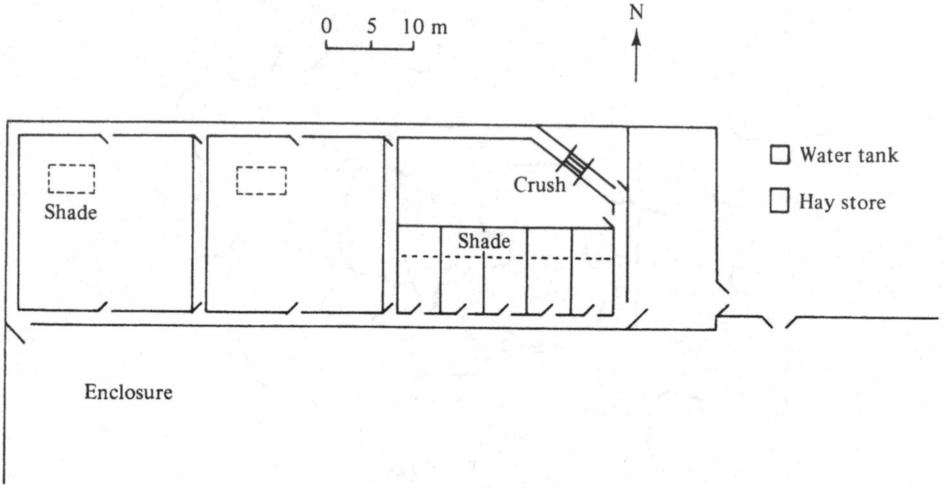

shade, and a row of five smaller isolation or calving pens, the latter with about half their floorspace concreted and shaded. Each of the larger pens was separated by a 1.5 m passageway, to prevent damaging fights between adjacent males, and all enclosed within a perimeter track of the same width. The main pens had north and south gates the width of the passage, and these could be opened into the pen or swung outwards to block the passageway when moving groups or single oryx in a clockwise direction. A movable screen allowed the passageway to be blocked at any point. A 3 m gateway led into the enclosure from the perimeter track, while the pens had 4 m access gates, with safety zone, and a nearby gate led into the enclosure.

Fig. 5.3. Vegetation types of oryx enclosure and surroundings at Yalooni.
a, Pan sump with sandy substrate; *Prosopis cineraria* woodland, with *Acacia ehrenbergiana* clumps, little ground cover.
b, Mosaic of sandy patches and stony substrate; *A. ehrenbergiana* with *A. tortilis* on harder areas. Locally, inside and outside enclosure, abundant tussock grasses.
c, Stony, flat plain with pavement surface; occasional *A. tortilis* and good herb cover where protected from gazelles and camels.
d, Stony, gently inclined slopes of runoff area, low bluffs and sharp, looser stone; occasional *A. tortilis*, scattered perennial shrubs.

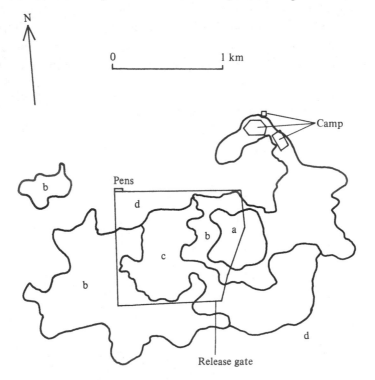

In the north-east side, the passageway angled and narrowed to pass through a handling crush. It had front and rear timber doors suspended from greased, overhead rails while the sides were plywood and tapered to ground level to hold an adult oryx but prevent it from turning or jumping out.

The large pens had no fixed feed troughs. With new or nervous oryx, the food could be thrown through the gate, or plastic bowls pushed in. With calmer animals, troughs with the feed lifted to 0.5 m above ground were used. A water trough in each pen then replaced plastic bowls. Each small pen had a feed trough inserted in the northern fence. All food and water was hand-carried to the pens. With only one group of oryx, the animals were soon trained to be herded morning and evening around the perimeter track, through the crush and into the pen which had not been used immediately beforehand, allowing the vacated one to be cleaned. Only one newly arrived oryx occupied one of the small pens (5.2.5), which were then used only infrequently, for seriously sick oryx or calves being hand-reared (5.2.8).

The enclosure of 100 ha (Fig. 5.3) was deemed large enough to allow a herd of 10–15 oryx scope for some natural social behaviour and the opportunity to exploit their environment. It was sited to include as great a range of vegetation types as Yalooni offered, and to be within easy reach of camp without the resulting human activity disturbing any oryx inside. Patterns of oryx use showed that each vegetation type met specific requirements (5.2.6).

5.2.5   *Management in pens*

Although all the oryx received in Oman were familiar with captivity, the different local conditions required every effort to minimise the chances of injuries from fences, posts or gateways from animals taking fright. This was attempted through very gradual and progressive training of the animals once they were calm in their new quarters.

On arrival at Yalooni the crate of the oryx judged tamest by the accompanying keeper was opened first into a 20 × 20 m pen (Fig. 5.2). The others of the group were then released into the same pen as quickly as possible. The only exception to this was the largest male of group 1, Jadib, who was released into a 5 m × 10 m pen because he had been quarantined separately from the other four animals at SDWAP. However, because he was so excited and was developing shoulder abrasions from the fence, he was united with the others after one day. A few bales of oat hay from SDWAP accompanied the first group as a familiar diet, but this was discontinued because of the readiness with which newly arrived animals ate the local lucerne and hay.

Until the oryx were calm in the presence of people outside the pen, food and water were put in plastic bowls close to the gates for easy replenishment with little chance of injury to the animals. The group was observed constantly by day at this stage from a distance which did not apparently disturb the oryx. When the animals stood to watch people, the gates of their pen were left open and lucerne dropped across the thresholds and into the perimeter passageway. The oryx were then left alone to find their own way out of the pen and to explore the passageway, learning that by progressing around it they returned to the same pen by a different gate. Progress of their exploration was monitored from their footprints in the sand of the passageway. After only 3–4 days like this, the oryx would leave their pen as a group by one gate if three people approached through the opposite gate. In this way the group could be herded around the perimeter and into the other pen, allowing the original one to be cleaned of faeces and feed refusals. After another 2 days this transfer between pens became a routine morning and evening exercise, and the oryx were used to passing through the crush. Thus, if examination or veterinary treatment became suddenly necessary, any animal could be caught efficiently with little stress. At this stage also, the plastic feed and water bowls were replaced by larger metal troughs which allowed more animals to feed or drink alongside and also reduced wastage.

The length of time spent in the pens by each oryx varied considerably (Table 5.1). The factors determining this depended less on the animals' habituation to their new conditions, which was complete within a month, than to the availability of the enclosure, its grazing condition and the composition of animals already there. Thus, group 1 was released into the empty and unused enclosure after 1.9 months. Hadya of group 2 was held for only 2 days because of her highly excited state from the moment of arrival. She was at risk of injuring herself severely and of alarming the others of her group. Museba and Rahaima were released into the developing herd 1 next because they had been reared together at Brownesville Zoo. After 1 month for their gradual absorption into the herd (see below), the final male and female, Nafis and Sajba, were added. Sajba had damaged her scapula in the pens while Hadya was being separated for release and her partial recovery was added reason for release last into the enclosure.

Lubtar and Farida, the founders of herd 2, were let into the enclosure 2.5 months after herd 1 had been released into the desert. This release into the enclosure was timed for the oryx to benefit from the new grazing resulting from rain while the enclosure was empty. Farida was also due to calve in less than 1 month and it was clearly advantageous for her not to do so in the pens. Because Farida had been mated in the pens, it had been necessary then to remove the other two males, Mustafan and the older Kadil (Table 5.1), to the second pen to avoid serious fighting with Lubtar. Lubtar and Kadil were both socially mature males and would never have co-existed peaceably in one herd. As Kadil was old, 113 months on arrival, small-bodied with very worn horns from rubbing fences, he could not be released into herd 2. Mustafan was held back in the pens for 20.4 months in anticipation of the arrival of group 4 from the USA, but was released before its arrival into herd 2 which, by then, had increased to five through the addition of two calves from Farida and one young female returned from herd 1.

Zeena and Hababa of group 4 were held for only 5.9 months in the pens. As both were excitable and had arrived in mid July, when the extreme heat caused them to hyperventilate easily, their initial management proceeded slowly. Mustarayha, once weaned from her mother in SDWAP, had occupied a display pen, resulting in complete lack of fear towards people. This made her unfit for release into the desert. Khalifa was moved through the pens quickly because he had come from a free-running herd in a large enclosure in Jordan (3.8), was young enough to be absorbed relatively easily into herd 2, and the release of this herd from the enclosure was scheduled for only 3 months after his arrival in Oman.

5.2.6   *Release into the enclosure*

The main vegetation types of the 100 ha enclosure are shown in Fig. 5.3. On release from the pens, single oryx or groups usually ran close to the west fence, often 'play-prancing' at intervals. Grazing started very shortly unless there were resident oryx already there, which immediately galloped to the newcomers. In this case there followed a period of intense circling by pairs of oryx, of either the same or different sex, and mutual sniffing under the tail. Aggression occurred exclusively between pairs of the same sex. Resident adult females tossed and lunged at newcomers, who tried to move to the edge of the herd past other oryx who were trying to sniff them. Adult males circled each other before suddenly turning parallel to interlock their horns and make sharp stabs to escape their opponent's lock while delivering a blow to his neck. When stabbing, both oryx pushed against one another, and might even fall to their knees in the process. The pushing rather than the stabbing probably showed which was the stronger. After 2–3 min of combat the weaker male would break off and be chased out of the herd.

After this initial meeting, new oryx might leave the herd to make at least one complete circuit of the enclosure. The level of aggression from the residents was always less when they subsequently returned to the herd, and the integration of females was promoted by the herding efforts of the dominant male. New males might move out of the herd as soon as the dominant male turned towards them over the next two months. For the next few days and weeks the new oryx were progressively tolerated, which could be followed in the frequency and type of interactions (Table 5.2). After the addition of these two, the percentage of all threats directed to them declined with time, as did the overall rate of interactions in the herd. Neither gave many threats to other animals because of their low status.

All oryx released into the enclosure most commonly ate grasses. Initially, some individuals tried to browse on the green leaf of *Acacia tortilis* and *A. ehrenbergiana*, and *Prosopis cineraria*, but the combination of thorns and the inability of the animal's broad, heavy-lipped muzzle to remove leaf efficiently soon led to their avoidance. Areas with good grass cover were soon discovered and heavily used, and the species seen to be avoided was principally the common and aromatic forb *Pulicaria undulata*. As the grass cover declined, the oryx stripped the relatively soft outer leaf layers from the abundant wiry *Crotalaria aegyptiaca*.

From their earliest days in the enclosure, the first oryx preferred to rest under one or several adjacent *A. tortilis* that provided deep shade, situated on the central plain of the enclosure with good visibility. Subsequently, some trees on this plain and at the periphery of the *Prosopis* woodland were much used. The animals of herd 2 may have been influenced by the presence of droppings from herd 1 at such key sites. In winter the oryx rested in the woodland and sandy depressions, excavating pits 20–25 cm deep in the leeside of the sand-mounds around each *A. ehrenbergiana*, to obtain protection from cold winds. Such observations showed that the oryx learnt very quickly to exploit the various vegetation types of the enclosure, and that its resources

Table 5.2. *The percentage of all threats in the herd given and received by two oryx, Nafis and Sajba, at 0, 3 and 7 weeks after introduction to herd 1 in the enclosure. Total herd size 10*

			Threats at successive weeks (%)					
		Age at introduction	Given			Received		
Oryx	Sex	(months)	0	3	7	0	3	7
Nafis	m	31	2.8	4.7	1.4	32.7	17.1	19.1
Sajba	f	23	2.8	2.8	1.4	29.1	18.0	11.4
Threats/hour			14.4	11.7	5.6			
Hours of observation			8.0	7.5	5.4			

m, male; f, female.

were utilised in different ways depending on season. Apart from some food and water (5.2.7), the enclosure met all the oryx's basic needs of habitat structure.

5.2.7 *Provision and consumption of food and water*

When the first four oryx of herd 1 were released into the enclosure from the pens in 1980, the available grazing was dry and sparse, for the last heavy rain had probably fallen in 1977–8 and other herbivores had been excluded for only four months since completion of the perimeter fence. Thus provision of the same feeds of air-dried lucerne and coarse hay continued once a day in the evening, unless a morning feed was desirable for management purposes such as close approach for immobilisation or in preparation for release (5.3.2).

The quantities of food given were calculated on the basis of the amounts eaten *ad libitum* in the pens. The aim was to compensate for the shortfall in natural grazing so that the animals remained in a condition adequate for reproduction, rather than being able to deposit obvious amounts of fat. Table 5.3 shows the average amounts of lucerne and hay consumed daily per adult oryx equivalent over 21 months leading up to herd 1's release. The adult equivalent is calculated for each animal as 0 between birth and 6 months of age, 0.5 between 6 and 18 months, and thereafter 1.0. The oryx received a total of 1.62 kg/day per adult, with lucerne and hay in the ratio 2.7:1.0. Using laboratory determinations of nitrogen in these feeds, intakes are compared also in Table 5.3 with figures for the fringe-eared oryx under penned

Table 5.3. *Feed consumption by herd 1 in the enclosure. Fringe-eared data from Stanley Price (1985a)*

	Hay	Lucerne	Total
Average kg/day per adult	0.44	1.18	1.62
s.D. kg/day per adult	±0.16	±0.24	±0.28
Average crude protein, %	10.4	16.0	
Average water content, %	3.4	16.7	

	Arabian oryx	Fringe-eared oryx
Crude protein of diet, %	14.5	7.4–12.5
Dry matter intake as % body-weight	2.31	2.04–2.39
Dry matter intake (g/kg $W^{0.75}$ per day)	66.9	66.8–75.1

conditions (Stanley Price, 1985*a*). The intakes of the Arabian oryx lie within the ranges of the other species, indicating that the Oman animals in the enclosure were receiving a maintenance level ration in addition to any gain from their grazing.

Water was provided every evening from a trough, and its consumption measured by depression of the level, and the surface area of the water. Attempts to allow the oryx access to water less than every day were abandoned because their efforts to remove the trough's cover were liable to lead to injury or horn damage. Monthly average consumption shows the strong seasonal influence of changing air temperature (Table 5.4). By comparison, grazing fringe-eared oryx drank 3.6 l/head per day in a mean ambient temperature of about 30 °C (Stanley Price, 1985*b*).

5.2.8   *Hand-rearing of calves*

Although hand-rearing had no part in the ideal preparation of oryx for release into the wild, it became unavoidable when the first two calves born in Oman were abandoned by their mothers. The circumstances and causes are discussed later (7.2.6).

The first calf, Selma, was rescued after wandering, bleating, through the enclosure for 72 h in temperatures exceeding 40 °C in May 1980, by which time it was evident that she had been abandoned. She took to a bottle willingly, and appeared to suffer no dehydration. She was then reared on a soya-based powdered milk, which was an inadequate substitute for the protein-rich,

Table 5.4. *Water consumption by herd 1 in the enclosure*

	l/adult per day		
	1980	1981	1982
January		1.8	2.5
February		1.7	
March		2.8	
April	5.5	3.9	
May	3.2	4.2	
June	4.1	5.0	
July	3.4	4.7	
August	4.8	4.8	
September	2.9	3.3	
October	3.2	3.4	
November	3.5	3.0	
December	2.4	2.5	

concentrated milk of oryx. Although this left her permanently stunted, her ability to breed was unimpaired (7.2). At 6 weeks old she was taken for walks in the enclosure from the pens, with a slightly older, orphaned gazelle, who stimulated feeding by imitation. At 3 months, she encountered the herd, which included her mother. As weaning progressed, Selma gradually became more absorbed into the herd, although likely to disregard herding efforts by the male and to gallop up to the author, on whom she was heavily imprinted. This bond fortunately broke with her first calf.

When Alaga was born and abandoned in 1981, the rearing process was changed to reduce the likelihood of human imprinting. The powdered milk was changed to one based on cow's milk. After being taken in, Alaga developed pneumonia, which was successfully treated with antibiotics, respiratory stimulants and manual stimulation of her breathing for 36 h. When healthy at 1 month, she and another orphaned gazelle were taken to a shelter constructed for them in the enclosure. Its height allowed entry by Selma, who was showing an interest in the calf, but provided protection from the adults who might have been aggressive. The shelter was only 20 m from the oryx troughs and Alaga always had her bottle here, often while her mother fed nearby. Her weaning was simpler and earlier, as she watched and copied the oryx who were accustomed to her, and she became a herd member with no residual influence of her upbringing apparent in her behaviour.

5.2.9 *Identification and naming*

Each oryx arriving in Oman came with a studbook number, an ear notch denoting its number at its breeding establishment, and a name. Most of the names were Arabic and were retained, with the rangers quickly learning to recognise each oryx when the animals were in the pens or enclosure. Correct identification at this stage helped accurate reporting on the oryx when they were in the desert under observation from greater distances. Although the face patterning was very consistent between individuals, there were many differences in horn length and shape, their wear patterns or breakages. Body shape, the extent and intensity of black markings, especially along the flank, and the overall impression of the oryx, all helped identification.

Calves born in Oman were not ear-notched, largely to avoid any risk of mismothering after handling. However, each calf was named by the rangers once its sex had been established either by seeing how it was licked by its mother or from its urinating stance. Names chosen by the rangers for some significance in their culture were more easily remembered by them. They were often chosen with a family theme and to sound harmonious, for example Selma *bint* (daughter of) Salama, or Fadi *bin* (son of) Hadya.

5.3   Release into the desert

5.3.1   *Assessing readiness and timing of release*

The two prime criteria to be met before oryx were considered ready for the open desert were that they should appear to be fully acclimatised, and that the herd should be integrated.

In the absence of clear signs of distress at extremes of temperature, such as panting or hyperventilation, the degree of physiological acclimatisation could not be judged through observation alone. In any case, deficiencies in the amounts of food and water available could be made good by management after release, and the most obvious clues to oryx adaptation and response to the conditions of Oman were their coat colour and texture. If the normal sequence of seasonal changes were shown by immigrant animals (2.4.1), it was assumed that complementary physiological adaptations were occurring. However, the group 2 oryx, arriving in Oman on 8 December 1980 from California, did not attain their very light summer coat in the following summer. Throughout the summer of 1981 they remained distinctly less white and more furry than the group 1 animals in the same herd. Thus, complete adaptation of coat may require considerable periods of exposure to Oman's climate. This is clearly relevant to the optimum season for receiving animals from overseas and for releasing animals into the desert.

The gradual development of the integrated herd was watched as all individuals moved round the enclosure in a characteristic spatial arrangement (6.2). Their activity patterns became increasingly synchronised (6.2). At the same time, the rate of interaction in the grazing herd dropped to less than one per hour, and the interactions themselves were very low intensity adjustments of attitude or a mild stare. Such herds have a constant and unambiguous dominance hierarchy involving all males and females above 3–4 months old. The experience of herds 1 and 2 has shown that the top and second males are dominant to all females, but that the hierarchy of adult females is then dominant to all other mature and immature males. It is likely that in oryx herds in which additions are only through birth, dominance position largely depends on age.

These two criteria determined the minimum period before any oryx were released into the desert. The actual times spent in the enclosure (Table 5.1) also reflected the time taken for the integrated herd to be assembled from more than one group of immigrant animals. Release was also subject to ecological considerations. In view of the possibility of the released herd scattering and running, it was felt best to avoid mid summer, with the increased chance of heat-stress, and also mid winter, when day-length, and hence the time available for searching for lost animals, was less. Thus, spring and autumn were judged to be the best seasons, and the herds were released on 31 January 1982 and 4 April 1984. The state of grazing was the final factor determining release season. The chances of wide dispersal by the released animals were reduced if grazing conditions ensured that the oryx were still dependent on a supplement which they would associate only with the enclosure. Thus, paradoxically, optimum conditions for the animals' growth and performance may be detrimental to a successful release and initial establishment.

5.3.2   *Preparations for release*

Preparations for the release of a herd fell into two categories, namely staff training and logistics, and oryx management.

During the period in the pens and enclosure the project's rangers learned to recognise each oryx and observe their behaviour from on foot at distances of 30 m, then sending their observations to the camp as a report over a handheld radio. Some further training was required, because surveillance after release was by vehicle-borne patrol of two rangers. Exercises were held in intercepting strange vehicles and obtaining relevant information from the driver, while reporting back to camp by radio. Before the first release each ranger was competent at radio-location, homing-in systematically either on a transmitter beacon or a radio-collared camel. These exercises ensured that staff, vehicles ar·l equipment were ready to meet any eventuality as the oryx left the enclosure.

Radio-collars were fitted to two males and two females of herd 1 after immobilisation two months before release (Woodford *et al.*, 1988). Three oryx of herd 2 were fitted with collars while they were still in the pens and could be handled in the crush without being immobilised.

In the 3–4 weeks before a release, the schedule of feeding was altered so that the total ration was split into equal morning and evening feeds. The feed troughs were moved progressively every few days across the enclosure towards the release gate (Fig. 5.3), so that for a week the animals were accustomed to feed less than 20 m from the gate. On the early morning of the release day the troughs were moved to less than 20 m outside the gate. As the animals could then see their feed, the opening of the gate was delayed for an hour after their normal feeding time, and a quiet exit from the enclosure was assured.

The sex and age composition of the two herds released are shown in Table 5.5. Selma was released twice because her hand-rearing left her less integrated into herd 1. After 7 months in the desert with this herd, during which she was mated, she was returned to the enclosure because of her persistent wandering around camp. While herd 1 contained only two animals born in Oman, there were four in herd 2 (excluding Selma) because of Farida's two calves born in her 23.7 months in the enclosure (Table 5.1) and the contribution of Selma's first two calves.

The two immigrant animals, Kadil and Mustarayha, who were unsuitable for release into the desert (5.2.5), were kept in the pens for 42 and 20 months, respectively, until transferred to the Oman Breeding Centre for Endangered Species in Muscat.

5.3.3   *The first month in the desert*

The response of oryx released into the desert is described in detail in Chapters 6 and 7. The initial reactions of the two released herds are described briefly here in the context of their management.

Herd 1 walked through the release gate as a tight group and fed at its troughs for 5 min before moving off, again bunched, to explore the nearest woodland, frequently pausing to look around. Within 20 min it had completed a loop and was back at the troughs. Later in the morning the herd followed the south fence to its south-west corner and then turned south to spend the middle of the day on a ridge 2 km from the release gate. The herd returned to feed that evening at the troughs, and in their first night outside they

Table 5.5. *Compositions of herds 1 and 2 at release*

	Herd 1		Herd 2	
	Name	Age (mo.)	Name	Age (mo.)
Males				
	Jadib	45	Lubtar	80
	Nafis	41	Mustafan	68
	Museba	35	Khalifa	14
	Hamid	33	Sulayman[a]	5
Females				
	Salama	43	Farida	74
	Talama	34	Selma	47
	Rahaima	34	Zeena	27
	Hadya	29	Hababa	23
	Selma[a]	21	Mundassa[a]	22
	Alaga[a]	8	Kateeba[a]	14
			Mafrooda[a]	12

mo., months.
[a] Born in enclosure.

circumnavigated the enclosure to visit the pens from the outside. The pens were an important focus for their movements, for the herd visited them 10 times in the first 12 days after release, but then only another four times in the following 16 days. This reduction was partly promoted by gradually moving the feed troughs from the release gate, so that after 24 days they were 4 km from the gate.

At release, water was available every day from a trough that was left inside the enclosure. After 3 days, one oryx became separated from the herd and paced inside the fence parallel to the others outside. Once he was outside, the water bowser was relocated outside and the gate shut. After 6 days, water was only available every other day to reduce water dependence, and this continued until heavy rain fell on day 25.

The initial pattern of movement was usually modest by day, but tracks and radio signals showed that the herd was mobile at night. On the night of day 16, all set out from the pens, but two, the dominant male and female, returned after 1 km. The remainder proceeded down a track into the north-west wind for 7 km, turning when they were 1 km downwind from a Harasis camp. All were near the pens the following morning, having walked a total of 16 km, during which it was evident they had grazed, explored sand-mounds, particular trees, etc. Excluding this excursion and any others not observed, the herd's range at the end of its first month was 13 km². Release caused no apparent social changes, and only the dominant male squat-defaecated in conspicuous sites as inside the enclosure.

Herd 1 was 40 km from Yalooni when herd 2 was released. The latter's behaviour on release was also restrained. Within minutes of walking through the gate, animals were grazing on the grass remaining from rain the previous year, and the dominant male squat-defaecated near the south-east corner of the fence. The herd shaded on this first morning only 200 m from the gate. Although Selma knew the local area well from having been released earlier in herd 1, the herd bull actively prevented exploration, herding oryx back towards the release gate and away from the pens, in marked contrast to the behaviour of herd 1. On the third morning, the herd was found returning towards Yalooni, having walked a 10 km loop up to 4 km from the gate. On day 8, the youngest and least integrated oryx, Khalifa, was tracked up to 10 km from Yalooni. He drank when water was offered, and was easily persuaded to walk back with three people herding 20 m behind.

This herd's supplements were halved on release because of the natural grazing, and after a week feed was given only once a day, encouraging wider ranging. The water bowser was towed to 70 m outside the enclosure on release day. Over its first 26 days this herd had a known range of 32 km².

5.4   Management after release

5.4.1   *Aims*

The released herds of oryx had the option of moving to wherever they wished, despite their partial dependence on supplementary feed and water, which the animals knew were available only in the release area. Management effort at this stage was directed to ensuring the security and wellbeing of the animals. More specifically, the main priority after the release of each herd was the prevention of any possible illegal hunting. Levels of surveillance allowed almost continuous records of herd location, with efficient and early reporting of potential problems such as abnormal behaviour or separations from the herd.

5.4.2   *Security and communications*

A ranger patrol comprised two rangers in a vehicle with high-frequency radio communication to Yalooni camp, other patrol cars and even the Muscat base (Plate 4*b*). The car on duty also carried receiving equipment for radio-tracking oryx. Every patrol was fully equipped for a standard period of 3 days and 3 nights on duty if its herd was more than 15 km from Yalooni. A minimum of a complete spare patrol was always on stand-by duty in camp for back-up as needed.

In addition to specific observations (5.4.5), the rangers watched a herd from a range of 0.1–0.5 km, depending on conditions. Each morning they routinely reported the location of the herd. If a single animal had separated or the herd had clearly moved far during the night, a second patrol might assist. The patrol was also responsible for knowing where any Harasis were living in the area and ensuring that these people were advised of the herd's location and any special conditions such as lying-out calves. Any unknown vehicles in the vicinity of oryx were stopped and the driver asked to detour the herd if necessary.

5.4.3   *Influence of patrols on oryx*

In the first months following the release of herd 1, patrols were readily permitted to influence the direction of herd movement by strategic positioning of the vehicle or by turning and herding gently. Thus, when herd 1 descended the Huqf escarpment (4.2) and then set off across a large salt flat, with no productive area at the far side, four cars were used to reverse this move. As evidence accumulated of the oryx's 'good sense' and sensitivity to ecological change, this policy evolved to one of alert *laissez faire*, unless there was specific cause for intervention. For example, patrols were instructed to herd any oryx from gazelle carcasses because of the risk of botulism (7.3.2).

In general, intervention was rare and oryx quickly habituated to the patrol's presence. On occasion, because of the daily association of car and herd, lost or separated animals used the more conspicuous car as an indication of herd location, so that appropriate siting of the vehicle could be a useful management tool.

The patrol cars had a distinctive appearance with, in general, a slight negative association for the oryx because of their herding function. As the oryx were attracted to the vehicle which brought their feed, only a pick-up was used for this purpose. Any station-wagon was regarded as completely neutral, and allowed closest approach for purposes such as filming or immobilisation. Thus, enforcement of different vehicle use for specific purposes aided management.

5.4.4   *Locating oryx*

Efficient monitoring of the oryx required the rangers to locate a herd or individual oryx quickly. As the rangers rarely found their herd visible in the morning from their campsite or near the herd's previous evening's location, they had to find the herd on most mornings. There were four main techniques.

(1) *Radio-tracking*. All the rangers were proficient at radio-location. A directional, double-Yagi antennae null-peak system 20 m above ground in camp had a maximum reception range of 10 km. A car-based aerial could rarely receive more than 7 km even when the transmitter was in prime condition. As an oryx could walk this distance in under 2 h, and much greater nightly moves were made (6.3), radio-tracking was of most use for almost continuous tracking of herds when near Yalooni immediately after release, or for detecting whether a herd had started to travel after sunset. Diminishing signals in the evening at least showed the rangers in which direction to start the following morning.

Whether transmitting or not, the radio-collars served another purpose: to any outsider seeing a collared animal, it was obvious that the oryx was protected or under surveillance in some incomprehensible way. This was reason for not harming it, and the project encouraged the rumour generated by the rangers that, if a collared oryx was molested, a signal automatically flashed to Yalooni!

(2) *Visiting known haunts of the herd, group or individuals*. This was a relatively effortless option for a motorised patrol, but unless the herd's movements were very local, it resulted in random searching

with only a very low success rate. It was discouraged because of its inefficiency compared to the next method.

(3) *Searching for tracks.* The patrol returned to the last place in which the missing oryx or group had been seen or before any split in the group had occurred. The Harasis could do this by virtue of their very precise site recall. It might take 2 h to identify the relevant tracks from the confusion in an area heavily used by the herd, but this was not wasted time. Once found, the correct sets of tracks would always lead to the missing animal or group, and the rangers never had to admit defeat with a set of prints.

(4) *Systematic searching.* If a specific animal or group had not been seen for some days or longer, the last sighting's location would almost certainly be empty of oryx, and finding the tracks leaving the area would be fruitless because of the profuse wind-blown hoofprints, all of apparently the same age. Under these circumstances, systematic searching was justified and exploited the oryx's conspicuousness (2.4.2). This was most effective under a low sun and with a search pattern to minimise the area scanned when looking into the sun's glare. With the sun behind the observer, sighting distance was limited only by undulations in terrain, for a singly, exposed oryx was easily visible in the morning or evening at 5 km with binoculars, and 2–3 km to the unaided eye. The chance of overlooking oryx decreased with group size. This enabled large areas to be searched efficiently.

5.4.5   *Research observations*

Various research and monitoring observations were made on the oryx at different intervals (Table 5.6). Most of the daily information was provided by rangers, who also reported on oryx growth and condition, sexual activity and any uncommon behaviour.

In addition to the purposes in Table 5.6, the rangers' motivation was improved by requiring them to watch for an end-point such as the time the last oryx entered the shade each day. Some observations were educational through being comparative: for example, the rangers saw for themselves the oryx's superior water economy and their indifference to sandstorms, compared to the reactions of goats or gazelles.

These observations were collated as a computerised database on each oryx, which yielded a monthly summary of herd structure, ages and adult equivalents, breeding performance, calves born and expected calving dates. The unstructured observations on behaviour and performance assisted management by showing patterns of individual behaviour such as in pre-calving separations or wandering.

5.4.6 *Veterinary care*

Once in Oman, the oryx received no prophylactic veterinary treatment. Their faeces never indicated more than very light levels of gut parasitism. The number of engorged ticks showed that animals newly released into the enclosure became fairly heavily infested, but they adapted to this challenge for tick loads soon dropped and stayed low on all oryx in adequate condition. In March–May, the oryx attracted small biting flies around their

Table 5.6. *Research and monitoring observations on wild oryx*

Observation	Purpose
Daily	
1. Herd location	Determination of home-range, ranging pattern, range expansion
2. Times into and out of shade	Oryx response to climate, and evidence of water conservation
3. Location of lying-out calves	Extra surveillance to locate calf and confirm its wellbeing, allowing quick response if accidental separation from mother
4. Amounts of feed and water	Assess need for supplement, to monitor demand for water or independence
Opportunistically	
1. Behavioural observations on activity, dominance relation-ships leadership, separation	To understand social organis-ation and dynamics
2. Feeding observations	To develop list of plants eaten
3. Size and condition	For qualitative comparison of growth rates and performance
4. Sexual activity	Prediction of calving dates, estimation of gestation length, studbook entries
5. Extent of isolation of mother and newborn calf from herd	Semi-quantitative indicator of rate of calf development and quality
At intervals	
1. Faecal nitrogen level	Indicator of diet quality
2. Diet nitrogen level	Indication of diet protein level
3. Faecal parasite ova check	Assess level of gut parasitism

heads, but they seemed merely irritating. There was no indication of infectious disease passing through the herds, and there was no disease transfer from the local livestock.

Professional veterinary assistance was far from Yalooni, and a veterinarian could only be flown in for an emergency. One reason for close observation of all oryx was to detect signs of problems early. Apart from those occasions when veterinarians visited for oryx that subsequently died (7.3), their assistance was required only once for immobilisation to fit radio-collars, once to treat a groin abscess on a yearling, and once to treat a female who initially appeared to be having difficulty in aborting a seven-month foetus.

5.4.7   *Provision of food and water*

In view of the advantage of releasing oryx from the enclosure at times when natural grazing was sparse outside (5.3.1), supplementary feeding had to continue either until rain fell and there was new grazing or the oryx moved from the release area to better conditions. The level of supplementary lucerne and hay was again kept to the minimum to maintain the animals' condition, and this also ensured that the herds grazed with natural activity patterns to increase their intakes.

The periods and extent of supplementation are described in Chapter 6. Initially, both herds' feed was supplied in troughs which were moved

Fig. 5.4. Locations of water troughs of herds 1 and 2. Dates are those of arrival of trough at new site.

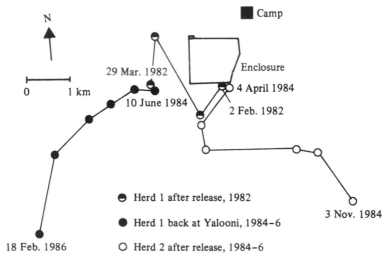

N

0   1 km

■ Camp

29 Mar. 1982

Enclosure

4 April 1984

10 June 1984

2 Feb. 1982

18 Feb. 1986

3 Nov. 1984

◗ Herd 1 after release, 1982

● Herd 1 back at Yalooni, 1984–6

○ Herd 2 after release, 1984–6

gradually away from the release area to encourage range expansion (Fig. 5.4). Each herd had its own water point following their first meeting (6.5). The feed troughs followed, advanced or ranged on either side of the water troughs by up to 3–4 km. In August 1985 it was apparent that the herds' daily foraging ranges were being limited by their evening return to the troughs. From then on any supplement was taken to where the herd was each afternoon. This was up to 20 km from Yalooni for some groups of herd 2 oryx, and 15 km for herd 1.

The quantity of supplement given was adjusted frequently from the animals' condition and assessments of the quantity and quality of grazing. Supplementation was self-terminating either because the oryx spurned lucerne and hay when their grazing intake was adequate, or because they had made a major change of range following rain (6.3.2). The corollary of this was that if a herd returned to the release area from far away it was either for water or because the grazing of their entire range was inadequate. Under these circumstances there was no merit in withholding extra feed or water.

Experience showed that while oryx might expect to find grazing or artificial feed anywhere, they had a specific site attachment for water sources. This seems adaptive to natural conditions where water sources are scarce and distant, requiring accurate navigation to locate them on rare visits. Thus, moves of the water bowser were more conservative than those of the feed troughs and it was only moved either as the oryx watched, or to where the herd was.

5.4.8   Predation

Although the density of potential predators of adult oryx was low (4.6), goats were taken by wolves from Harasis flocks during the course of the project. While there was no known predation attempt on adult oryx, the policy in such an event was that the oryx must learn to cope with predators. Only in the event of an individual predator appearing not to have been deterred from further attacks, either by the oryx's behaviour or any management action, would direct steps be taken to eliminate the predator. Obviously, this policy should change to one of total non-interference as oryx numbers increase.

Lying-out calves were at greater risk of predation because of resident caracals and brown-necked ravens, and wintering golden eagles. Ravens killed one calf aged 1.7 months (7.3.2). Large numbers were attracted to the shade, water and food of Yalooni, where they persistently mobbed and pecked any recumbent gazelle or oryx.

5.5   Discussion

The pattern of management of oryx in the pens and their slow release into the enclosure to develop integrated herds show that management procedures were conservative. This was necessary because of the value of each animal on arrival at Yalooni, the absolute rarity of the species, and the low probability of the US sources meeting requests for replacements for animals killed or injured while in captivity in Oman. On the other hand, the integrated herds would have been developed in less time had it been possible to hold animals earmarked for Oman together in the USA. During such a phase the correct balance of ages and sexes could have been achieved, and any obviously unsuitable individuals discarded. Integration in Oman would also have been promoted by reducing the times between arrivals of groups from the USA. These intervals were the consequence of positive blue-tongue tests, subsequent selection of further animals for Oman from a limited population and the initiation of new export applications. Allowance must be made for such factors in the planning and scheduling of any re-introduction scheme involving international transport of founding animals.

From the 18 oryx received in Oman, two were unfit for even the first stages of preparation for release into the desert. The male, Kadil, was stunted, with worn horns, and paced fence-lines like a typical zoo animal. The female, Mustarayha, was completely tame. Both animals' traits made them so obviously unsuitable for release into the wild that they should have been selected out while in the USA. This emphasises the need for normal behaviour in captive-bred animals destined for release.

The preparation procedures for re-introduction used in Oman resulted in the release of habituated and partially dependent animals. This guaranteed that the released oryx could be watched from close range. Wildness and independence then developed at rates determined by the animals as they moved further from the release area in response to changing conditions (6.3). The animals' habituation to humans certainly made them more susceptible to illegal hunting, but their security on the Jiddat-al-Harasis seems assured (8.7). This allowed close monitoring of the released animals' responses and reaction to their ancestral environment, to the great benefit of the project (Chapters 6 and 7).

6

Oryx herds in their natural environment

6.1   Introduction

This chapter is the first of two which describe the response and performance of the re-introduced oryx in the desert. While Chapter 7 considers individual performance, this chapter is concerned with events following release of the two herds described in Chapter 5. The observations described were collected as part of the deliberate monitoring and surveillance of released oryx. They provide insights into the extent to which the re-introduced, and for the most part captive-bred, animals showed progressive adaptation and increasingly skilled exploitation of their ancestral habitat. Thus, this chapter has four main themes or lines of evidence. Most emphasis is on the first released herd because of its longer period in the desert and experience of a wider range of conditions.

The first theme is concerned with the movement patterns of released oryx, to assess whether their pre-release management was correct, to show how the animals' range expanded and to identify the causes of expansion. In the second section, relationships between the two herds are examined, as well as social dynamics within herds under re-introduction conditions. The occurrences of splits and separations from herds are summarised, with an analysis of their causes and significance.

The third theme deals exclusively with the released oryx's water economy, because of its central role in the viability of the re-introduction. It attempts to show whether captive-bred oryx, which have never known water-deprivation, can become water independent under the desert climate and presents some observations on how this may be achieved.

Finally, the fourth theme explores the oryx's perception of their natural environment, using specific observations to see how the oryx performed the feats of the first three themes. A case-history involving one oryx is then included. The behaviour of this animal was so different from that of the herds, in some ways resembling the translocation and release of unprepared animals, that it vindicated the management methods used for herds of oryx.

6.2   The herd as a cohesive unit

The re-introduction technique was underpinned by the assumption that the assembled integrated herd would have a long-term stability, and that this would allow effective monitoring of the animals following release into the desert. In this section the ways in which group cohesion is maintained are described.

An Arabian oryx herd moves around as a structured group. When grazing, females are closely followed by their dependent calves, but looser associations of females with mature offspring are also seen more often than would occur by chance. Females without young calves, and immature animals, tend to lead a grazing herd, while the dominant male is always at the rear, immediately behind any female with an active, newborn calf (2.4.4). Mature but sub-ordinate males range further from the herd, usually on its flanks, but their movements are influenced by the sexual state of the females and the proximity of the dominant male. When the herd is making a major journey, it adopts a closely spaced single file, with females and young calves at the rear with the herd male. An adult but subordinate male leads, often 100 m ahead of the next oryx.

When the herd is feeding in distinct patches of vegetation, such as a *haylah* of 1–2 ha, the animals are relatively closely packed, and the most frequent distance between animals is 5–10 m (Plate 3*b*). An adult female in the vanguard initiates the move out of the *haylah* to the next one. She starts to move in her desired direction, then stops and turns to look over her shoulder at the rest of the herd without altering the direction of her body axis. As other animals lift their heads and see this invitation posture they start to follow her, and the dominant male may actively herd the rear oryx in this direction. The male may attempt to prevent this movement by running to the front and vigorously turning the female back. When grazing is sparser on open, less-vegetated plains or ridges, or is more localised in the minor drainage lines between ridges, the oryx feed more spread out. A herd of 12–15 oryx may cover 0.5 km × 0.5 km, but individuals may wander 0.5 km from their nearest neighbour.

By day, and probably also by night, herd cohesion is maintained by visual contact. Even in the gently undulating country of the Jiddat-al-Harasis, an oryx standing 1 m at the shoulder can lose sight of the rest of the herd between low ridges in minutes. This animal immediately mounts a ridge to locate its fellows, and will then usually move to join them. A few observations suggest that subordinate males move to position themselves between the main body of the herd and outlying feeding females, so that a chain of visual contact is established, reducing the chance of separations. Separation is obviously a greater risk at night, although the Arabian oryx is highly reflective in moonlight. On nights with fog, soft 'moos', audible to the human ear over 30 m, probably maintain herd cohesion, although many short-term separations occur on such nights (6.6.1).

Two other, interconnected aspects of oryx behaviour promote herd cohesion. The first is the natural tendency of oryx to aggregate from their feeding dispersion both when starting to shade in the middle of the day and again at sunset. The average distance between shading oryx is 1–2 m, and as many animals as possible share the shade of one tree (Plate 2*b*). The herd male and adult females share the main tree while lower-ranking animals and all other adult males use as few satellite trees as necessary. The density of *Acacia tortilis* on the Jiddah means that these groups will rarely be more than 30 m from the main group. That this aggregation is not purely a response to the availability of adequate shade is shown by the similar concentration when the animals are basking in cooler weather (Plate 2*a*). On winter evenings the herd aggregates and each animal lies before dark on the sheltered side of a grass tussock or sand-mound in a *haylah*, with lying oryx commonly less than 10 m apart.

Fig. 6.1. Average time spent shading in 2-week periods and average maximum shade temperature (°C). Regression fitted to all points where temperature $\geqslant 27.0$ °C. Point ▲, 16–30 April 1983, omitted from regression (see the text).

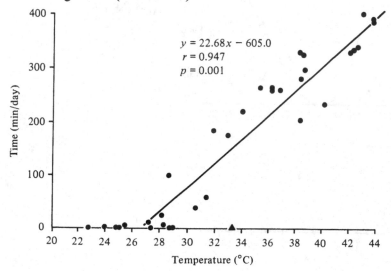

$y = 22.68x - 605.0$
$r = 0.947$
$p = 0.001$

Such behaviour shows how the climatic extremes of the desert (4.4) influence activity patterns through their effect on temperature regulation. The time spent shading each day is directly related to maximum shade temperature once it exceeds 27 °C (Fig. 6.1). During the hottest months the oryx shade for nearly 7 h each day, requiring long night-time grazing bouts. The anomalous point on the graph for 16–30 April 1983 has an explanation: the oryx were not

Fig. 6.2. Herd 2 in the morning and afternoon of 25 September 1986. Percentage of observations in four activities over 10-min periods. Activity of each of 12 oryx was recorded every minute, so that each percentage is based on 120 data points. Percentages not shown for 'other' activities, which in all but 2/46 periods accounted for ⩽ 10%.

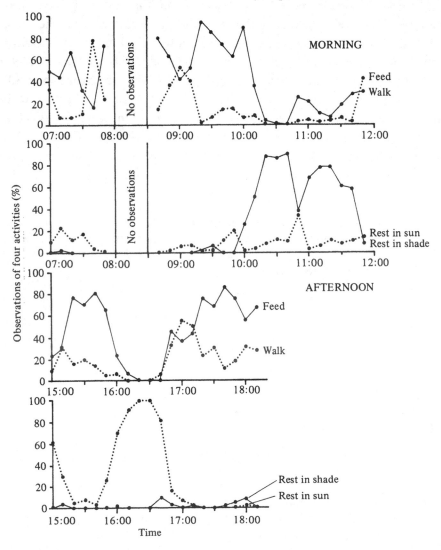

shading, because clouds of biting midges had invaded the Jiddah from the Huqf following heavy rain. As drinking water was abundant, the oryx preferred to rest on open ridges to reduce the insect challenge, despite the lack of shade.

On winter nights the oryx are inactive to prevent heat loss, and must consequently feed and bask during the daylight hours. The dictates of climate and thermo-regulation suggest that activity amongst the oryx of a herd will be highly synchronised (Fig. 6.2). On a typical late summer day, with some cloud, a herd fed, walked and basked in the sun in early morning. A mid-morning feeding bout terminated abruptly around 10:00 h as most oryx shaded. As morning observations ended, resting was being replaced by walking and feeding. In the afternoon, the herd had two distinct feeding periods separated by a period of basking in the sun, rather than in the shade as in the morning. The percentages of records for each activity tend to be very high or very low, giving a clear picture of synchronised activity and, consequently, a cohesive herd.

These observations on the spatial organisation of the herd, the active communication between herd members and the co-ordination of activity indicate that each oryx has firm bonds to its herd. This is the basis for using the herd as the fundamental unit in range establishment by re-introduced oryx.

6.3   Herd 1 range development

Although the chosen release area was good habitat for oryx (4.6.2), the pattern of movements in and away from it were likely to be instructive. The ever-present patrols sent daily or twice-daily fixes of herd location to Yalooni. Precise and accurate position reports were possible for any area once rangers and managers were equally familiar with its landmarks, often the tallest or most distinctive trees. These were mapped in each new oryx area and linked to the network of accurately surveyed markers left by seismic operations. All locations were coded on the UTM 1 km × 1 km grid system over the entire area and entered into the computer program RANGE (Appendix 4), which yielded the results of the following sections. These demonstrate the oryx learning to exploit their novel habitat through phased exploration, while responding to the desert's ecological stimuli, and progressively developing a collective memory of the Jiddat-al-Harasis.

6.3.1   *Range areas and occupancy*

The total area used by herd 1 from release at the end of January 1982 to 31 July 1986, covered 1932 km² (Fig. 6.3). It was not evenly situated around the release area, lying predominantly to the east and north-east of Yalooni, and with another major portion to the south-west. Thus, the range of herd 1 lay entirely within the ecological unit of the Jiddat-al-Harasis (Fig. 4.2). This total range area was composed of 12 individual ranges, each of which was used for a period following a major change of location. Ranges overlapped when the same area was used at different times. (Fig. 6.4a–d and Table 6.1). Occupation of both ranges 4 and 6 was divided into two (A and B) around the date on which rain fell. For each range, section A lay totally within section B. Range 10 was divided into three disjunct sections for purposes of area calculation only, for they were not occupied sequentially.

The smallest range covered only 35 km² (Table 6.1) and the largest 365 km² but the remaining 10 ranges all lay between 100 and 300 km². The highest figure, for range 12, is not strictly comparable for it was still occupied at the end of the study period, and it is an overestimate because it is the sum of the ranges of several subgroups within the herd (6.7). Occupancy of each range

Fig. 6.3. Herd 1's total range area from release (31 January 1982) to 31 July 1986. Routes joining areas occupied are shown in Fig. 6.5. Broken lines are roads as in Fig. 4.6.

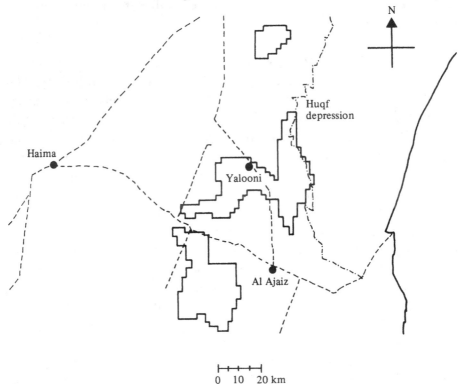

0   10   20 km

Table 6.1. *Range areas of herd 1, between release and 31 July 1986, with duration of occupation, area and cumulative area, the proportion of each range that was new, the period since rain at the start of each range occupation and the average distance from Yalooni of locations within each range*

	Range occupation			Cum.	New	Months	Distance from
Range	From/to (day/month/year)	Period (months)	Area (km²)	area (km²)	range (%)	since rain	Yalooni (km)
1	31/01/82 31/08/82	7.0	181	181	100	0[a]	5
2	1/09/82 3/01/83	4.1	269	450	100	0.5	20
3	4/01/83 11/03/83	2.2	162	612	100	0.3	57
4	12/03/83 28/04/83	1.5	288	779	58	—	23
4A	12/03/83 30/03/83	0.6	126	—	12	2.5	
4B	1/04/83 28/04/83	0.9	291	—	53	0	
5	29/04/83 2/06/83	1.1	226	994	95	1.0	24
6	3/06/83 1/10/83	3.9	256	1025	12	—	22
6A	3/06/83 9/08/83	2.2	98	—	0	2.1	
6B	10/08/83 1/10/83	1.7	256	—	11	0	
7	2/10/83 15/03/84	5.4	269	1294	100	1.7	61
8	16/03/84 13/06/84	2.9	101	1372	77	7.2	45
9	14/06/84 6/10/84	3.7	35	1376	11	10.1	3
10	7/10/84 18/12/84	2.4	265	1491	43	13.9	46
11	19/12/84 20/06/86	18.0	220	1614	56	16.3	7
12	21/06/86 31/07/86[b]	(1.3)	365	1932	87	0.2	58

Cum., cumulative.

[a] Rain fell 3 weeks after herd was released from the enclosure into range 1.

[b] End of present analysis.

Fig. 6.4. Range development by herd 1, 31 January 1982 to 31 July 1986: (*a*) ranges 1, 2, 3; (*b*) ranges 4A, 4B, 5; (*c*) ranges 6A, 6B, 7, 8, 9; (*d*) ranges 10A, 10B, 10C, 11, 12.

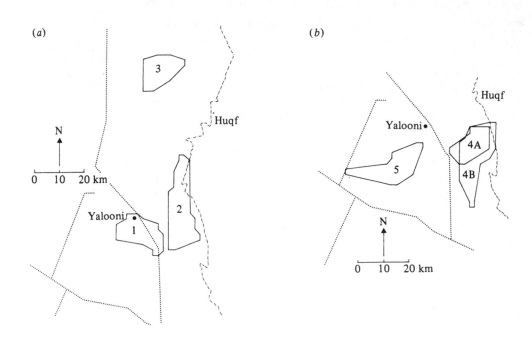

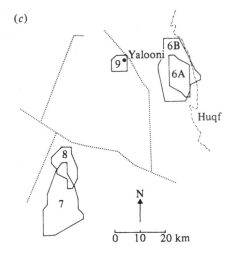

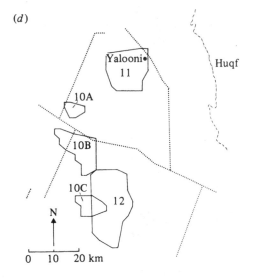

varied at between 1.1 and 18.0 months, with the majority falling between 2
and 6 months. The average distance in each range from the release area varied
at between 3 and 61 km, with the lowest values for the three periods of use of
the release area and its vicinity (ranges 1, 9 and 11).

6.3.2   *Factors influencing movement between ranges*

Movement into new ranges was most obviously caused by rainfall
through its effect of stimulating fresh grazing. Rain fell on the oryx in ranges
1, 4 and 6. Rain in other areas caused the herd to move into ranges 2, 3, 5,
7 and 12 between 0.2 and 1.7 months after the rain (Table 6.1). Consequently,
87–100 % of each of these ranges was used by the herd for the first time (Table
6.1).

The delays between rainfall and the herd moving from a non-rain area to
one with very recent rain (0.5 months from range 1 to 2; 0.3 months from
range 2 to 3; 0.2 months from range 11 to 12) suggest that the oryx did not
respond to lightning or to the appearance of black cloud. Observations at the
time of most journeys showed the herd moving when the wind changed to
blow from the direction of the rainfall. Comparing the occurrence of the wind
heading nearest to each journey's heading (i) in the period between rainfall

Table 6.2. *Journeys to new ranges following rainfall: heading of journey with
closest wind direction in 48 h before journey, and the frequency of this
heading in the period between rainfall and 48 h of the journey. Wind records
from Yalooni, taken at 4 h intervals*

Ranges		Days between rain and journey	Heading of journey	Nearest wind heading in preceding 48 h	Frequency of wind of same heading between rainfall and 48 h before journey
From	To				
1	2	14	80°	110°	1/103
2	3	9	345°	350°	2/56
4	5		No wind records		
6	7 (a)	57	225°	200°	4/203[a]
	7 (b)	63	200°	200°	3/32[b]
11	12	9	200°	200°	3/64

[a] Wind records for 28 days only before journey
[b] Period between journeys (a) and (b) used, not since rainfall. Sections (a) and (b) refer
to angled journey (Fig. 6.5c,d).

and 48 h before each journey, and (ii) in the 48 h preceding the journey (Table 6.2), the general premise is confirmed that the oryx detected and moved towards rainfall areas only after receiving information on an appropriate wind.

When range changes were made not in response to rain within the previous 2 months (into ranges 4A, 6A, 8, 9, 10 and 11), the herd tended to return to previously used areas, for the percentages of these ranges that were new lay between 0 and 77 % (Table 6.1). Various factors, acting either singly or in conjunction were responsible for this type of move.

The northern range 3 had fresh but sparse grazing on the herd's arrival in mid winter, but the area was treeless (4.5). By early March the air temperature had risen (monthly mean maximum 29.1 °C) and the oryx were starting to shade (Fig. 6.1). The inadequacy of the few trees for shade and the decreasing grazing caused the move to range 4.

Nearly one month after the rain of 1 April 1983, herd 1 left range 4B. As this rainfall was very widespread and heavy (Fig. 4.5), the herd was likely to find grazing wherever it went. An added proximate factor in their departure was clearly the activity of biting midges, which had invaded the Jiddah, while rainwater stood, from the seepages along the Huqf escarpment.

The return from range 5 into a known area that became range 6A showed the operation of several factors. Following the rain of 1 April 1983, all surface water had dried by mid May and, although grazing remained abundant, temperatures were rising. Unless the herd returned to the release area for water, it faced the prospect of a summer without drinking. Having used a total area since the last rain of 517 km^2 (ranges 4B and 5), the herd returned to the largest *haylah* of their eastern range for the summer. This provided good shade; grass samples contained 5% crude protein and annual forbs in their diet up to 14 % crude protein, both on a dry-matter basis. The water contents of their food plants were high (6.9.4). Under these conditions and with the herd shading for long periods by day, for the June and July maximum shade temperatures averaged 42.9 and 42.8 °C, respectively, the oryx were able to meet their nutritional requirements and obtain adequate water from their green diet, while their activity patterns facilitated water conservation. This situation changed totally with further rainfall on 10 August 1983, when the herd expanded into range 6B.

The circumstances of the herd's return to the release area (range 8 to 9) in June 1984 were monitored closely because the herd had then been away from Yalooni for 22 months and there had been no rain for 10 months. The protein level in the grazing of range 8 had dropped to below the maintenance standard of 4% (Table 6.3), while food moisture content had decreased. Large daily wind runs and high air temperatures led to very desiccating conditions in June (Table 6.3). By contrast, Yalooni provided better shade and shelter from such winds, had a water supply and a crop of very nutritious pods of *Acacia tortilis* and *Prosopis cineraria*. However, the natural grazing at Yalooni was also dry and had been utilised by herd 2 since the latter's release 2.3 months before herd 1's return. Herd 1 was, therefore, supplemented again 13 days after its return.

When air temperature and wind run had dropped in October (Table 6.3), the herd left Yalooni to occupy the large range 10, which overlapped ranges 5, 7 and 8 and contained a new sector. Without rain, the grass showed some regrowth due to fog moisture but it was no more nutritious than in the summer, even if more palatable (Table 6.3). After 2.5 months on such a marginal diet the herd again returned to Yalooni (range 11), and remained there for 18 months until the rain of June 1986, which drew the oryx into range 12.

6.3.3 *Factors determining occupancy and density*

The ranges were occupied under conditions of recent rain to great dryness (Table 6.1). Ranges occupied in response to rain less than 1 month

Table 6.3. *Crude protein (CP) and water contents of herd 1 diet in 1984, with climate and herd location at the same time*

		April	May	June	July	/	October	November
Foods								
Stipagrostis	%CP av.	5.25		3.41				3.97
sp.	S.D.	0.66		0.35				0.37
Prosopis	%CP av.				12.6			
cineraria	S.D.				1.5			
Acacia	%CP av.				15.3			
tortilis	S.D.				1.1			
Climate								
Av. max shade temp (°C)		39.9	40.3	43.6	40.7	/	35.7	31.2
Wind run (km/day)		326	431	388	541	/	230	209
Evaporative demand (cm/day)		15.5	14.6	21.5	14.5	/	11.6	8.8
Herd 1 location		Range 8		Range 9 (Yalooni)			Range 10	
Diet		Grazing only		Grazing and Supplements			Grazing only	

av., average; S.D., standard deviation.

previously are neither consistently larger nor smaller than those used in drier periods (Table 6.4). The spread in range area is less than in length of occupancy, and they are not related ($r = 0.007$, not significant). Thus, after oryx have moved into an area, they do not explore and continuously add to the area used. There are other determinants of range extent.

Rain often falls on the Jiddah as localised storms (4.4). The large move made by the herd from range 2 to 3 was in response to such a storm, and the area of range 3, 162 km², was seen on the ground to coincide with the limits of fresh grazing. However, even if on this occasion storm area determined the area used, the rainfall of 1 April 1983 covered the whole Jiddah. Following it, the oryx used ranges 4B and, a month later, 5. Their areas were only 291 and 226 km², respectively. This suggests that even when new grazing is superabundant, a herd limits the area used to less than 300 km².

Oryx density as km²/oryx gives further clues (Table 6.4). The average is 22.0 (± 7.6) km²/oryx, but, in general, the lowest densities occur in ranges

Table 6.4. *Ranges grouped by dryness, with range area, oryx density and average daily distance moved by herd 1*

Dryness	Range	Range area (km²)	Oryx density (km²/oryx)	Distance moved (km/day)
Occupation < 1 mo. after rain	1[a]	181	19.0	2.8
	2	269	29.9	4.3
	3	162	17.0	4.3
	4B	291	30.6	6.5
	6B	256	28.4	6.2
	12[b]	365	30.4	5.7
Occupation 1.0–< 6 mo. after rain	4A	126	13.3	4.7
	5	226	23.8	4.9
	6A	98	9.8	3.1
	7	269	26.9	3.6
Dry grazing or regrowth, > 6 mo. after rain	8	101	10.1	2.9
	10	265	25.2	4.8
Very dry, with supplementation	9	35	2.9	1.3
	11	220	19.1	3.0

mo., months.
[a] Rain fell 3 weeks after herd released into range 1.
[b] Range 12 was still occupied at end of study period.

occupied after rain. As these are times of greatest food abundance, the densities do not reflect the areas needed to meet food requirements. Thirty km²/oryx seems a lower limit, a density below which the herd does not live. This density coincides with a maximum range size of 300 km² for a herd of 10 adult oryx. No reasons for this limitation can be proven, but 300 km² may be the maximum area which a herd can know intimately enough over the time-span of range occupancy to be familiar with its geography and resources for efficient utilisation.

The highest density occurred when the herd was being supplemented daily at a fixed point, in range 9. In range 11 rations were taken to the herd wherever it grazed, resulting in a lower density. When the herd was on poor grazing in ranges 8 and 10, it lived at 10.1 and 25.2 km²/oryx. Although diet quality may have been approximately the same in both (Table 6.3) and some of each range was common, range 10 comprised four areas, resulting in the lower density (Fig. 6.4). Ranges 4A and 6A had high densities. Range 4A was occupied as conditions dried out, and range 6A was an area chosen by the oryx to spend their first summer far from Yalooni without water to drink (6.3.2). Their strategy was to settle in a small area, chosen from a very large area used after the last rain (ranges 4B and 5), with lush grazing that would have been adequate for the summer had further rain not fallen and taken the herd into range 6B.

The movements of herd 1 show that range area is influenced at least by rainfall area, presumed limits of familiarity, and the interaction of diet quality, water conservation and response to high air temperature. A more

Table 6.5. *Journeys of herd 1 between ranges*

Ranges				
From	To	No. days	km	km/day
1	2	2	20	10.0
2	3	6	66	11.0
3	4	3	82	27.3
4	5	4	51	12.7
5	6	2	46	23.0
6	7	6	103	17.2
7	8	1	6	6.0
8	9	2	40	20.0
9	10	2	40	20.0
10	11	1	28	28.0
11	12	3	58	19.3

quantitative analysis would require data from the Jiddah on an impractical scale. The data available confirm the importance of rainfall in stimulating oryx to move (2.3), and this is borne out by the effects of rain falling on the oryx themselves. This happened twice, in ranges 4 and 6. The data (Table 6.4) for 4A and 6A refer to the period before rain, and 4B and 6B from rainfall until the next major moves into ranges 5 and 7, respectively (Fig. 6.4). In these

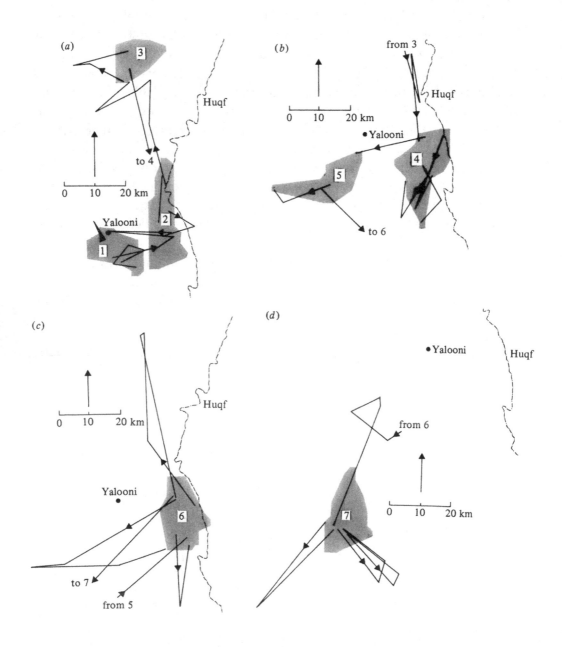

cases, rain stimulated increases in the daily distances moved by 138 and 200%, and the area used per oryx increased by 230 and 290%. The picture is consistent that rain falling on oryx stimulates them to travel more, presumably to explore the extent of the rainfall area. As the future, hot-summer range of 6A was explored by the oryx developing 4B after rain (Fig. 6.4; 6.3.2), the mobility caused by rain is adaptive.

Fig. 6.5. Journeys and forays between and from ranges by herd 1, 31 January 1982 to 31 July 1986: (*a*) ranges 1, 2, 3; (*b*) ranges 4, 5; (*c*) range 6; (*d*) range 7; (*e*) ranges 8, 9; (*f*) ranges 10A, B, C, 11; (*g*) range 12.

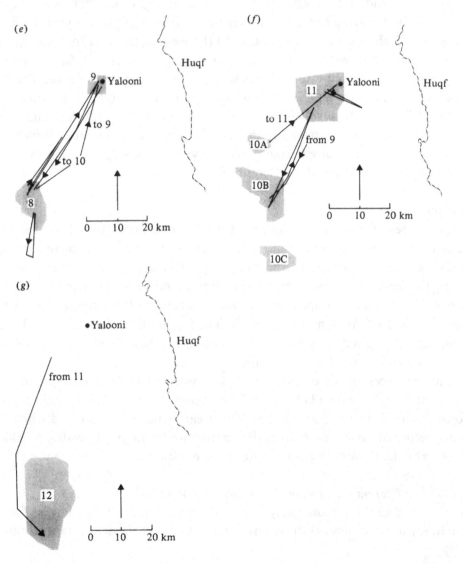

6.3.4   *Travels by the herd*

Two types of travel are recognised, which are distinguished on each of the range maps (Fig. 6.5*a–f*). The first type is the journey between successive range areas. The second comprise forays to beyond the current range area, followed by return to it.

Journeys

A total of 11 journeys between the 12 ranges varied between 6 and 103 km. The average distances travelled each day varied only between 6 and 28 km/ day (Table 6.5).

The herd did not necessarily move in a straight line between ranges, particularly if moving into an unfamiliar area. For example, the journey from range 2 to 3 shows the herd exploring to the west of the area that became its range 3, having moved from range 2 into wind after rain (Fig. 6.5*a*). When the herd vacated range 6 it returned first straight to range 5, where it presumably received information on the south wind, causing it to move in this direction to range 7 (Fig. 6.5*d*), which was entirely new (Table 6.1). When the herd left range 5 in June 1983 to return to the familiar range 6 for the hottest months, it did not take the direct route, but curved through new country, thus assessing the condition of a different area.

Forays

Two types of foray are recognised. Type 1 consists of a foray from the core area into hitherto unused country, followed by return to the current range. Striking examples are seen from range 7 (Fig. 6.5*d*). A type 2 foray occurs when the herd revisits a formerly used area, again returning to the current range. The longest examples of such forays occurred from range 6 when the herd made swift returns to ranges 3 and 5. All the forays discussed here involved all or a majority of the herd, and had to have lasted more than one day to be shown on the range maps.

The distances moved on forays varied between 11 and 106 km, spread over between 2 and 5 days (Table 6.6). The average daily distance walked on a foray varied between 5 and 37 km. The greatest distance occurred when the herd returned to range 6 from its excursion to range 3, walking 50 km overnight. Daily distances of 25–40 km were common.

6.3.5   *Factors influencing the occurrence of forays*

The three most likely factors determining the frequency of forays are likely to be air temperature, through its effect on activity patterns, rainfall and the consequences of the herd moving into a new range. These are not necessarily independent of one another.

Grouping the monthly occurrences of the 20 type 1 and 2 forays into four seasons of winter, spring, summer and autumn fails to produce adequate sample sizes in all categories (Table 6.7), but the impression is of few forays in summer, and even fewer in winter. This corresponds with management experience that the herd moves to a range which it will occupy for winter or summer, in the absence of any overriding ecological changes.

Forays to new areas (type 1) are more likely to occur within 30 days of rainfall (Table 6.8), and occur significantly more often within 30 days of a change of range. Obviously, rain is the major factor in causing range change, but the herd made three long forays soon after arrival in range 7 (Fig. 6.5*d*) despite the last rain having been almost two months before.

Table 6.6. *Forays by herd 1*

Range	Month	Foray type[a]	No. days	km
1	March	1	2	11
	August	1	4	28
2	September	2	2	38
	September	1	2	18
	November	1	2	31
3	January	1	4	34
4	April	1	3	30
	April	1	2	35
	April	1	2	22
5				
6	August	1	2	40
	August	2	4	91
	September	2	5	106
7	October	1	3	44
	October	1	2	73
	October	1	2	50
8	April	2	2	35
	May	2	3	27
	June	2	3	77
9				
10				
11	May	2	2	24
	February	2	3	69

[a] For explanation, see the text.

Totalling only seven, the type 2 forays are too few for any statistical testing (Table 6.8). However, the expected frequencies suggest that such returns to previously used areas, followed by return to the current range, are not related either to rainfall or a change of range area in the previous 30 days.

6.3.6 *The object of forays*

Forays had several purposes. Oryx returned to the release area to drink from range 2 and range 8 (Figs. 6.5a,e). From range 2 the herd also made its only descent of the Huqf escarpment. This and the three forays to the south of range 7 took the herd to the boundaries of the Jiddah ecological unit, and in each case the oryx returned. Generally, the type 1 forays allowed the animals to explore and to learn a new area. They occurred more commonly after recent rain as the oryx attempted to define the area of rainfall and superior grazing. Observations also suggested that such forays started on days which were cooler than average, usually because strong southerly breezes

Table 6.7. *The monthly incidence of forays of herd 1*

Month	D	J	F	M	A	M	J	J	A	S	O	N
No. forays	0	1	1	1	4	2	1	0	3	3	3	1
		Winter			Spring			Summer			Autumn	
Total		2			7			4			7	

Table 6.8. *The observed (Obs.) and expected (Exp.) frequencies of two types of foray by herd 1 within 30 days and after more than 30 days following 5 rainfalls and 11 changes of range, with χ^2 and probabilities for type 1 forays*

Foray type		Days after rain		Days after range change	
		< 30	> 30	< 30	> 30
1	Obs.	6	6	7	5
	Exp.	1.32	10.68	2.41	9.59
	χ^2		18.84[a]		10.94[b]
2	Obs.	2	6	2	5
	Exp.	0.88	7.12	1.61	6.39

Significance at [a]$P < 0.001$; [b]$P < 0.01$.

reduced the need for shading, and lush grazing allowed a shorter morning feeding bout.

6.3.7   *Movement within ranges*

The distances moved daily within a range by an oryx herd in the course of its normal activities are obviously under-estimated by the straight-line distances between successive days' locations. However, these distances are useful indicators of a herd's response to the prevailing local conditions (Table 6.4). The smallest distances occurred in range 9, when the herd was back at the release area and being heavily supplemented. The data suggest that daily distances were greater in ranges occupied less than 1 month after rain, and the effect of rainfall in ranges 4 and 6 are consistent with this (6.3.3).

Range area and length of occupancy are not related (6.3.3). The distance moved each day is not related to range area (Table 6.4, $r = 0.289$, not significant). However, daily distance is greater in ranges with short occupancy

Fig. 6.6. Months of occupancy and average daily distance walked by herd 1 in each range. Ranges where oryx supplemented (9 and 11) and range 12, occupied at the end of the study period, are excluded.

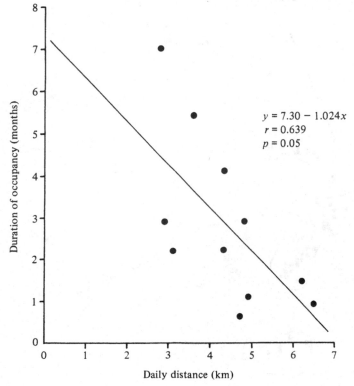

$y = 7.30 - 1.024x$
$r = 0.639$
$p = 0.05$

Daily distance (km)

Duration of occupancy (months)

and is smaller in ranges used for longer periods (Fig. 6.6). Thus, if oryx make large moves through their range, occupancy is likely to be short. Larger daily distance might reflect a scarcity or poor quality of grazing, which then prompts a change of range.

When grazing conditions were good, no pattern to the herd's daily movements was discernible. Under drier conditions the herd might move around exhibiting a form of rotational grazing, as in range 10 (Fig. 6.7). The herd's pattern of utilisation revealed four sectors. The most used area was Lefow North, which previously was range 8 and was re-occupied for 37 days

Fig. 6.7. Pattern of herd 1's movements in range 10. Each circle indicates a 1 km × 1 km cell occupied at least once.

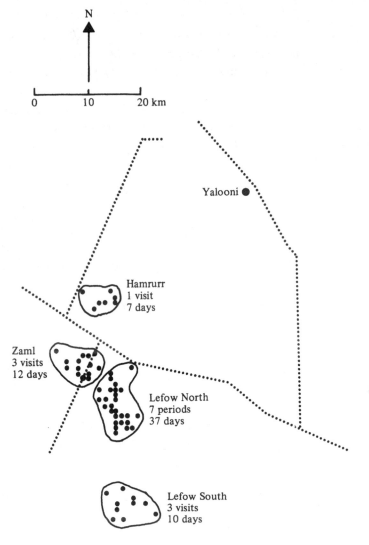

spread over seven periods. Between each of these periods, a visit was made to one of the other three areas. Lefow South was part of former range 7, and retained some dry grazing. The Zaml section was mostly new to the herd, and had good regrowth grazing on areas used by goats four to six months before, when the oryx had been in range 8. The Hamrurr area was also one of similar regrowth, and had formed the extreme western portion of range 5. The herd's pattern of movements, the scarcity of green grass and the protein content of 3.97 % in *Stipagrostis* sp. in November 1984 (Table 6.3) all indicated that the grazing was around or even below maintenance standard, precipitating the return to Yalooni and range 11.

6.4 Herd 2 range establishment

In marked contrast to herd 1's discrete, consecutively occupied ranges, herd 2's first four areas overlapped considerably, and their delimitation was necessarily more arbitrary (Fig. 6.8 and Table 6.9). This

Table 6.9. *Range areas of herd 2 between release and 31 July 1986, with duration of occupation, area and cumulative area, monthly range occupied, the proportion of each range that was new, the period since rain at the start of each range occupation, and the average distance from Yalooni of locations within each range*

| Range | Range occupation | | | Cum. area (km²) | New range (%) | Months since rain | Distance from Yalooni (km) |
	From/to (day/month/year)	Period (months)	Area (km²)				
1	4/04/84 16/11/84	7.4	79	79	100	12.1	3
2	17/11/84 23/08/85	9.1	166	169	54	19.5	6
3	24/08/85 25/02/86	6.1	178	249	45	28.7	9
4	26/02/86 13/06/86	3.5	288	347	34	34.8	16
5A	14/06/86 31/07/86[a]	(1.5)	205	552	100	0.1	39
5B	14/06/86 31/07/86[a]	(1.5)	297	849	100	0.1	70

cum., cumulative.

[a] End of present analysis; numbers in parentheses are not comparable with others in column 3.

different pattern was due to herd 2's release into the desert when there had
been no rain for 12 months, so that partial supplementation continued.

In the months following release, the herd moved southwards from the
release area, then to the south-east, where better grazing remained on a large,
sandy plain (Fig. 6.8a), before moving eastwards, but with continuing visits
to the release area (Fig. 6.8b). The easterly movement into range 2 was
initiated in November 1984 by a brisk east-south-east breeze following a
night of thick fog. The herd found an area, which had been heavily used by

Fig. 6.8. Range development by herd 2, 4 April 1984 to 31 July 1986:
(a) range 1; (b) range 2; (c) range 3; (d) ranges 4, 5A, 5B.

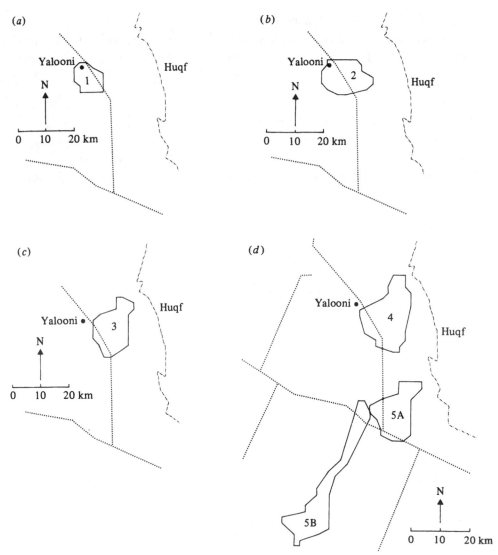

goats about 6 months previously, where regrowth of grass allowed a reduction in the herd's supplement. The increasing use made of this eastern portion of the total range (Fig. 6.8c) was encouraged by taking the herd's feed to it rather than requiring the oryx to walk westwards to their troughs (5.4.7), and lower winter temperatures allowed fewer returns to drink. Occupancy of this area also increased the separation between the two herds (6.5).

The starting date for use of range 4 was determined by the birth of the first of three calves, following which the herd became more fragmented (6.7). During the 3.5 months in range 4, the herd continued to range further from the release site, with an average distance from Yalooni of 16 km (Table 6.9). This was made possible by the presence of very light showers of rain on the herd, which produced a small flush of new grazing, and reduced the frequency of visits to water. Movement into range 5 was stimulated by heavy rain in mid June 1986. Unlike herd 1's divided ranges (4 and 6, Table 6.1), which were occupied by the whole herd before and after rain falling in them, the two portions of herd 2's range 5 were used simultaneously by two sections of the herd. The bulk of the herd, 12 oryx, used range 5A, while two females and their calves, aged 3 and 13 months, occupied range 5B. As these areas did not overlap, and this separation persisted long after the end of the period analysed, the areas cannot be merged for calculation of the total area used. At the end of this period, therefore, most of herd 2 had a total range area of 552 km² while the other four certainly knew some of range 5A by passing southwards through it, but occupied a further 297 km² of their own.

Compared to that of herd 1, the range development of herd 2 was more gradual in the face of dry conditions over the entire Jiddat-al-Harasis. Herd 2 was in the desert for 26 months before the rain of June 1986, and herd 1 was back in the Yalooni area for 22 of these months because of this dryness. Herd 2's range expansion was also more conservative, for no forays as described for herd 1 were observed.

The patterns of rainfall following release of the two herds account for the different rates of range development. After 26 months in the desert, herd 1 had occupied seven non-overlapping ranges, with a total area of 1294 km² up to 61 km from Yalooni. Some 26 months after its release, herd 2 had only two strictly discrete ranges, with an area of 347 km² at a maximum distance from Yalooni of 16 km. However, the immediate move of herd 2 to the rainfall area of June 1986 showed that its inherent response was not inferior to that of herd 1.

6.5   Relationships between herds

When herd 1 returned to Yalooni on the night of 12–13 June 1984 from range 8, after an absence from the release area of 22 months, it found that herd 2 had been occupying an area of 51 km^2 adjacent to the enclosure since its release two months previously. By 05:00 h on the 13th, the 24 oryx of the two herds had mixed, and the two herd males had fought, each presumably in defence of 'his' area. By 07:00 h the herds had separated into their original combinations, and remained like this through the 14th. They met and mixed again on the 15th, and this marked the start of a 6-week period in which the number and composition of groups were very fluid (Appendix 5).

The number of groups varied on any day between two and six. The most common size for the largest group was 14 oryx (21 observations), but this was biased because of the early and lasting move of one oryx from herd 2 to 1 (see below). However, groups of 14, 15 or 16 oryx out of the total of 24 for both herds accounted for the largest groups on 35 out of 49 days. Thus, the largest group was almost always identifiable as either herd 1 or 2 less a varying number of defectors or with some immigrants from the other herd. Groups containing 19 or more oryx of the 24 were recorded on only six well-spaced occasions (15, 29 June; 4, 9, 14, 23 July), during which many exchanges took place, and herd males interacted if present.

While the initial meeting and mixing of the herds was caused by inter-herd competition for the immediate vicinity of the release area, other factors certainly helped the fragmentation of the original herds and development of some of the smaller groups listed in Appendix 5. The groups of 1:0 and 0:1 were single females, Salama in herd 1 and Selma in herd 2 (the former the mother of the latter), exhibiting pre-calving separation from the herd. One male and a female separated for 3 days during the latter's oestrus, and other small groups split due to the activities of subdominant males attempting to drive away oestrous females. Some small groups were due to the separation of Haleema, attended by the herd male or with others, as she showed signs of the epileptic condition which was fatal in early August 1984 (7.3.2). The lasting exchanges between herds, described below, may have been based on bonds between particular individuals.

Group compositions stabilised by about 28 July. This occurred despite no significant increase in the distances between herds. On 31 July 1984 a calf was born into herd 2, and a week later one in herd 1. The two herds were then 4–5 km apart and the newborn calf in each reduced their ranging distances. At this stage, despite the recent fluxes in group composition, only four oryx (three males, one female) were not in their herd of before 13 June. Their behaviour is discussed below, but each oryx found in its new herd an animal of the same sex and almost identical age (Table 6.10). For three animals, this oryx in the new herd was closer in age than any in the herd which it had left.

Sulayman, born into herd 2, was associated with primarily herd 1 groups from 16 June, and never again returned to herd 2. In herd 1 he found Rahman, born in the same month, while his natal herd contained no young males other than Khalifa, who was 9 months older. Khalifa also forsook herd 2 from 14 July and found age peers in herd 1 but ultimately returned to herd 2.

Table 6.10. *Composition and ages of herds 1 and 2 in June 1984, showing which oryx changed herds in June and July*

	Herd 1			Herd 2	
	Age (months)	Oryx		Oryx	Age (months)
Males					
				Lubtar	84
	75	Jadib			
				Mustafan	72
	71	Nafis ————→			
	63	Hamid			
	19	Migdhaf			
		←————→		Khalifa	18
	16	Fadi			
	9	Rahman ←————		Sulayman	9
Females					
				Farida	78
	73	Salama			
	64	Talama			
	64	Rahaima			
	59	Hadya			
				Selma	51
	38	Alaga			
				Zeena	30
	29	Haleema			
				Hababa	27
		←————→		Mundassa	26
				Kateeba	18
				Mafrooda	16
	8	Askoora			

Mundassa joined herd 1 from 24 July, and accompanied it into new territory when it left for range 10 in October. She returned to Yalooni with the herd in December, and on 12 January 1985 she and a young male left the herd and located herd 2. He spent 11 days there before returning to herd 1 while Mundassa remained in her natal herd, which included her mother and two full sisters. During Mundassa's sojourn in herd 1 she conceived for the first time. By doing so, she avoided the likelihood of being mated by her father, Lubtar, who remained the breeding male in her natal herd 2.

The last exchange concerned Nafis, an adult but very sub-dominant male (7.4.2). His associations changed from predominantly herd 1 to herd 2 on 25 July 1984, but from August he commonly switched between the two herds either alone or with small groups. He finally rejoined herd 2 in March 1985 until dying in January 1986.

The 6 weeks following the initial meeting of the herds allowed each oryx to

Fig. 6.9. Range use by herd 2 only (solid line) in two periods, and herd 1 (broken line) and herd 2 in two periods.

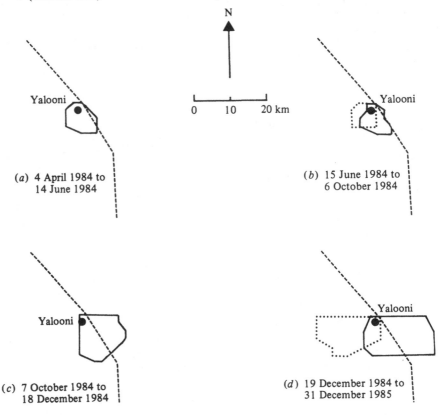

(a) 4 April 1984 to 14 June 1984

(b) 15 June 1984 to 6 October 1984

(c) 7 October 1984 to 18 December 1984

(d) 19 December 1984 to 31 December 1985

meet all others. As the herd compositions stabilised, only four changes had occurred, and of these only Sulayman and Nafis showed any permanence, although had the latter lived longer he might have oscillated between the herds when in the same general area. This level of lasting exchange was very low and testified to the strong bonds between individual oryx in herds assembled as herds 1 and 2 were in preparation for the desert. The sex ratio of 3:1 (male:female) in the oryx that did exchange suggests that females are less liable to leave their original herd, perhaps because female–female bonds are stronger than those between males.

Following the meeting of the herds in June 1984, the total area used by both herds around Yalooni was progressively partitioned into herd territories (Fig. 6.9). While herd 1 occupied its range 9 at Yalooni, it used an area of 35 km² and herd 2 41 km², of which 18 km² were common to both herds. The average daily distance between herds was 3.1 km.

After herd 1 left Yalooni again to use range 10, herd 2 expanded its range to the east and south (Fig. 6.9c), an area of 113 km², but it did not re-occupy the area vacated by the departed herd 1. Finally, when herd 1 returned to range 11, again close to Yalooni, both herds had larger ranges (134 and 173 km², respectively), but they overlapped only by 24 km² (Fig. 6.8d). The average distance between the herds had increased to 11.1 km.

These observations suggest that each herd had an almost exclusive herd territory after a period following their meeting. Although demarcated along their boundaries and throughout by the faecal piles of the dominant males of each, these males were rarely seen heading oryx away from the other herd's territory. The levels of exchange of individuals show that each herd was aware of the approximate position of the other when around Yalooni. However, the rarity of lasting exchanges suggests that oryx were merely avoiding those areas occupied by less familiar individuals. This development of herd territory around Yalooni was undoubtedly promoted by the deliberate siting of the water troughs (Fig. 5.4, p. 116) and then by taking the supplement to the herds while occupying the areas shown in Fig. 6.9 (5.4.7).

6.6 Separations from herds

The emphasis in this chapter on herd behaviour implies that each oryx remained in its original herd, apart from the periods following the meeting of herds 1 and 2. Although each oryx was in its herd for the majority of the time, separations, as distinct from exchanges between herds, occurred.

Separations are identified as short term or long term, depending on whether they extended over more than one day. Those lasting less than one day probably originated through a group from the herd starting to feed at night.

Such separations were commonly seen on mornings following a fog. The precipitated water stimulated grazing (6.9.4) at times when herd cohesion through visual contact was least possible. By daybreak the separated groups would be only a few kilometres from one another, invariably within the current herd range, and the groups were usually together again before the midday shading period.

The longer-term separations could be identified as deliberate or accidental, with the distinction made on the location or activity of the separated oryx. In general, the behaviour of accidentally separated animals showed that they would rather be with the herd.

Table 6.11. *Accidental separations from herd 1, 1982–5*

Year	No.	Group	Days	Distance from herd (km)	Reasons, comment
1982	1	1y	2	4	Hand-reared, liable to wander
	2	1y	2	3	As for (1)
1984	3	1f	2	14	Lost herd at night
	4	1f	5	60	Lost herd on move at night, after first signs of epilepsy, revisited former range
	5	1c	4	45	Failed to move with herd at night to Yalooni for drink
	6	1f	7	20	Epileptic, wandered from herd in night, revisited former range
	7	1m	73	40	Failed to leave Yalooni at night with herd for range 10; waited alone until herd returned
	8	3c	(17)[a]	40	Failed to leave Yalooni at night with herd for range 10; all confined in pens for vet. treatment of 1. On release, located herd in former range
	9	1f	4	35	Separated at night, revisited former range
1985	10	1f,1c	2	5	Left behind after herd visited water at night
	11	1f,1c	15	10	As for (10)
	12	1f,1c	8	10	Failed to detect herd visiting water at night, then revisited former range, 25 km distant

m, male; f, female; y, yearling; c, calf.
[a] Not comparable with other values in the column.

6.6.1   *Accidental separations*

Between release in January 1982 and the end of 1985 accidental separations from herd 1 totalled only 12, and all but one involved either a single animal or a female with dependent calf (Table 6.11). The shortest separations lasted only 2 days, with the longest 73 days by ex-herd male Jadib, who stayed in the release area when the herd left for range 10 in October 1984. Apart from the two separations of the hand-reared animals, all occurred when the main herd made a significant move at night. In separations 4 and 6 (Table 6.11), the young female concerned was exhibiting the first signs of intermittent epilepsy, and this may have affected her ability to maintain contact with the herd at night.

Excluding separations 1 and 2, all the separated animals rejoined the herd through their own efforts. The strategies for doing so were either to stay in the same place or search the current range for the herd (cases 3, 5, 7, 10, 11 and 12) or revisit a formerly used range in the expectation that this might be the herd's destination (cases 4, 6, 8, 9 and 12). Separations 10, 11 and 12 involved the same female and calf. After separation 11, these two were in the herd only 2 days before losing them again as they returned to water. After a further 8 days alone, they suddenly revisited range 6 25 km away. Finding no signs of oryx occupancy there, they returned to Yalooni and located the herd.

An accidentally separated oryx uses various tactics to relocate the herd:

(1) It checks for the presence of the herd in the immediate area by moving to a nearby ridge and scanning.
(2) The animal searches through the current range for tracks. Oryx have been watched following fresh tracks to the herd.
(3) If not searching, the separated oryx stands motionless on a prominent ridge, advertising itself by its whiteness.
(4) Some oryx tend to panic when alone from the herd, and when found by a patrol will approach the car because it is so strongly associated with the proximity of the herd.

6.6.2   *Deliberate separations*

Twenty-two deliberate separations, each lasting more than 1 day, were recorded from herd 1 over the same period (Table 6.12). Twenty separations lasted between 1 and 8 days. Of the other two, one lasted 65 days when a single male returned to the release area with the first symptoms of an *Actinobacillus* infection. His initial motive may have been to drink as the herd had then been 9 months without water in 1983 but he found a small herd 2 in the enclosure with whose male he interacted through the fence, and this may have caused him to become resident and territorial outside the enclosure

until his death. The other separation lasted 27 days and resulted in a change
in the herd's dominant male (6.8).

The distances between the separated oryx and their herd varied between 4
and 65 km, similar to the accidental separations. However, the composition
of the separated animal(s) differed slightly in that, of the 22 incidents, six

Table 6.12. *Deliberate separations from herd 1, 1982–5*

Year	No.	Group	Days	Distance from herd (km)	Comment
1982	1	1m, 1f	3	4	Oestrous female
	2	1f	2	15	Pre-calving separation
	3	1f, 1c	7	6	Post-calving separation: herd guided to 2 separated
	4	1m	2	25	Returned Yalooni for drink
1983	5	2m, 1f	(1)	10	Oestrous female; all guided back to herd
	6	1m	65	15	Return to Yalooni with first symptoms of fatal *Actinobacillus* infection
	7	1m, 1f, 1c	27	15	Forcible separation of female and newborn by male, preceding change of herd's dominant male
	8	1m	2	< 3	Expulsion from herd after demotion from dominance
1984	9	1m, 1f	5	28	Oestrous female
	10	3m, 1f, 1c	3	65	Revisit former range, concurrent with (9)
	11	1f	2	45	Return Yalooni for drink
	12	1f, 1c	3	45	Return Yalooni for drink
	13	1f	2	45	Return Yalooni for drink
	14	1m, 1f	4	3	Oestrous female
	15	1f, 1c	2	9	Revisit former range
	16	1f	2	25	Pre-calving visit to former range
	17	1f	3	10	As for (16)
	18	1f	8	16	As for (16)
	19	1f	6	65	As for (16)
	20	1f, 1c	6	15	Stayed in range 10B while herd revisited range 10C
1985	21	2m, 1f, 1y	2	20	Revisit former range
	22	1f, 1c	2	7	Post-calving separation with weak calf

m, male; f, female; y, yearling; c, calf.

separations involved more than a single oryx or a female and calf. The most common cause for the separation was pre-calving isolation (five incidents), followed by a return to the release area for a drink (four) or by a male and oestrous female consorting (four) away from the herd. In three cases a brief return to a former range was the motive, and the isolation of female and newborn calf accounted for two cases.

There are two further points concerning these separations.

(1) The male of separation 6 never returned to the herd, and management action guided the herd and separated animals together in separations 3 and 4 in 1982 and 1983. In all other cases the separated animals were left alone and returned and found the herd after absences of 2–27 days. This suggests either that the separating group knew that the herd was unlikely to leave the current range, or that the herd deliberately stayed in the same area until the group returned. Alternatively, both occurred in the absence of any change in conditions which would promote a move to a new range.

(2) Not all oryx were equally prone to departing deliberately from the herd. One female, Salama, was involved in all seven pre- and post-calving separations with various calves, in two out of four returns to Yalooni to drink, and in separation 20 (Table 6.12). Thus, many herd 1 oryx were never recorded as separated in this period.

6.6.3   *The significance of separations*

Separations did not occur equally through the years. Although action was taken on occasion in the first 2 years to guide separated animals back to the herd, this probably only shortened the period of separation and did not result in under-estimation of the number of separations of more than 1 day. So, the lack of accidental separations after release, other than those of the hand-reared animals, suggests that herd cohesion was effective as the oryx made their first explorations throughout 1982 and 1983. The accidental separations of 1984 occurred as the herd moved around after a long period without rain and fresh grazing and before returning to the release area from range 8 (6.3.2). Separations of both types were few in the second half of 1984 and through 1985, because the herd was occupying range 11 around Yalooni, and consequently moving little (Table 6.4).

As some oryx were more predisposed to leave the herd deliberately, separations were, overall, uncommon. The accidental and deliberate separations totalled 359 oryx-days, only 2.5% of the total oryx-days between the release of herd 1 and the end of 1985. A separation started on only 2.4% of all days, and all the animals of herd 1 were together for all but 21 % of the days. These data confirm the cohesiveness of a herd, which enabled efficient surveillance and monitoring in the desert.

6.7   Subgroups within current herd range

The separations described above all involved small numbers of oryx which had become separated from their herd either accidentally or deliberately. Starting in 1986, herds 1 and 2 both exhibited the development of subgroups for some reasons in common and others particular to each herd. These subgroups were characteristically of variable composition, usually lasted no more than a few days, with the animals staying within the current herd range and exhibiting no effort to locate the main herd.

Table 6.13. *Age (years) and sex structure of herds 1 and 2 in February 1986*

	Herd 1		Herd 2	
	Oryx	Age	Oryx	Age
Males				
	Jadib	7.8	Lubtar	8.6
	Hamid	6.8	Mustafan	7.6
	Migdhaf	3.2	Khalifa	3.1
	Fadi	3.0	London	1.0
	Sulayman	2.3	Mahyann	0.9
	Rahman	2.3	Antar	0.8
	Dtheeyab	1.6		
	Jumaa	1.2		
	Samir	1.1		
	Murshid	0.8		
	Mahmood	0.1		
Females				
	Salama	7.7	Farida	8.1
	Hadya	6.5	Selma	5.8
	Alaga	4.7	Zeena	4.1
	Askoora	2.2	Hababa	3.9
			Mundassa	3.7
			Kateeba	3.1
			Mafrooda	2.9
			Qurtaasa	1.6
			Tohayla	1.2
			Rasheeda	0.8
			Kasheeda	< 0.1

Development of subgroups started in herd 1 on 5 February 1986, and from then until the herd moved after rain to range 12 a total of 29 different subgroups was recorded, with a modal size of three oryx. Some of these groupings were associated with calving, as for the separations, but single males or groups of males were recorded for the first time.

Ecological conditions predisposed this fragmentation: range 11 contained almost no grazing, was small, close to the release area and hence familiar to all animals, and the daily supplement was taken to each group in the absence of rain. Every oryx must have known that there was no reason for the range to be vacated in preference for any other. Any group would have known its herd-fellows were within a few kilometres and, in any event, the water source provided a focal point. Thus, there was little penalty for oryx not being in one herd.

The sex and age structure of herd 1 (Table 6.13) also encouraged these subgroups. By February 1986, the sex ratio was highly biased in favour of males, and the first Oman-born males were becoming sexually and socially mature. At the same time, two of the breeding females were pregnant or had just calved but showed no *post-partum* oestrus, and after the third was mated on 25 April 1986, the dominant male had little motive for keeping these females together all the time.

The development of subgroups in herd 2 started with the birth of a calf on 26 February 1986, followed by births on 1 March 1986 and 20 March 1986. The new herd male (Mustafan) allowed two of these females to leave the herd separately with the former dominant male and a subadult. In addition, very light rain fell on herd 2's range in February and April, which caused isolated pockets of new grazing in the larger *haylahs*. Until the entire herd moved to the area of substantial rain on 13 June 1986, 49 different combinations of the 17 oryx were recorded, ranging in size from one to eight, but with groups of one to five equally frequent. The number of groups on any day varied between one and six, but on 60 out of 77 days there were three, four or five groups.

Some of the groups maintained their identities for up to 2 weeks as they occupied their own small area of grazing and failed to meet others on their occasional visits to water. In contrast to herd 1, the presence of only three males of more than three years old in herd 2 (Table 6.13) meant that some groups had no adult male. But, as with herd 1, the herd 2 members knew that all were in the same range, for the total area used by subgroups coincided almost exactly with that of the main herd.

6.8   Subgroups in the change of dominant male
In both herds 1 and 2, there was one recorded change of the dominant male. Both successions had features in common, involving the temporary separation of a small group from the herd at the instigation of the challenging male.

When herd 1 was in range 7 in November 1983, two females calved on the same night. Jadib, the 5.6 year-old herd male, could not consort with both (2.4.4). He, therefore, could not prevent 4.5 year-old Hamid from driving Salama and her calf away from the herd and most aggressively thwarting her attempts to return. Hamid herded the calf occasionally and started to squat-defaecate (2.2). The trio settled 15 km north of the herd in a hitherto-unoccupied area with good grazing. Hamid mated Salama successfully when the calf was 12 days old, but killed the calf at 27 days with repeated stabs and blows. Six days later he allowed Salama to head back towards the herd's location of 27 days earlier, where it still was. When the pair found the herd at night, Hamid, who was in better condition and may have had an advantage of surprise, ousted Jadib. The latter received superficial wounds but was progressively tolerated back into the herd as a subordinate over the next 10 days.

In herd 2, Lubtar had been dominant male since its development in the enclosure. When Mustafan was added 13 months later, he was 4.8 years and Lubtar 5.8 years old. Although Mustafan beat Lubtar in physical contest, Lubtar retained his superiority and Mustafan behaved as a typical subordinate male. From early June 1985 Mustafan made increasing efforts to prevent two females and their wild-born calves from returning to Lubtar and the main herd for the evening feed. On 25 June 1985, one of these females led the same group, with Mustafan following, to a *Prosopis* grove 5 km from the herd, to search for fallen pods. This marked the beginning of a period of 70 days during which Mustafan vigorously prevented the four from returning to the herd or in its last known direction. His herding reduced the frequency of visits to the joint herd's water and feed troughs (6.9.3).

Although 61 of the total 67 km² of Mustafan's group range were shared with Lubtar, the majority of Mustafan's locations were in a small area at the southern side of his range which had never been occupied by herd 2 (Fig. 6.10a). The overlap of these two ranges was exaggerated by the shared water source, for the average distance separating the two groups was 4.1 km. This ensured that the two groups were out of sight of one another for most of the time. The groups met on eight occasions in July at feed or water, but only once in August when the feed troughs were not used because feed was taken directly to each group.

After 70 days, on 2 September 1987, Mustafan's behaviour changed, as with Hamid (above), for he walked at the rear of his group, which he allowed to move northwards into the area then used by Lubtar's herd. The two groups met in the afternoon, at which Mustafan defeated Lubtar in a 10-min combat and ensured he did not return to within 0.5 km of the herd that day. By the following morning Mustafan and his same group of four were separate again from the main herd, but, for the next 23 days (Fig. 6.10*b*), its range was almost

Fig. 6.10. (*a* and *b*) Areas used by herd 2 and Mustafan's group over two periods.

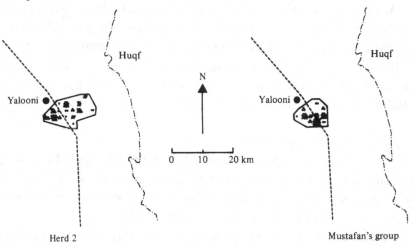

(*a*) 24 June 1985 to 2 September 1985

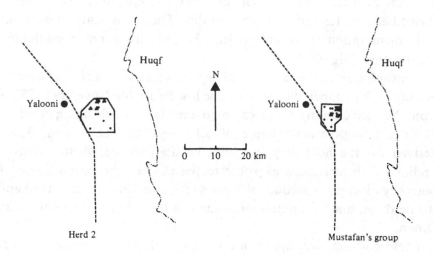

(*b*) 3 September 1985 to 25 September 1985

entirely enclosed by that of the herd. Consequently, the two groups were only 3.0 km apart on average, and on 3 days were plainly in view of one another. On 25 September the two groups merged after a subadult male had joined the smaller, and shortly afterwards Mustafan defeated Lubtar and became the undisputed dominant male.

6.9   Water economy

6.9.1   *Introduction*

The genus *Oryx* is renowned for its ability to occupy arid areas because of its low water requirements. Central to the popular mythology of the Arabian oryx is its survival for months in the desert without drinking. If the water economy of zoo-bred oryx had relaxed through selection or lack of experience in captivity, their re-introduction to the relatively waterless Jiddat-al-Harasis might have been compromised. The measurement of water consumption and circumstances of any water independence by the oryx were therefore of great concern to the project.

6.9.2   *Water consumption*

Water was provided *ad libitum* for all oryx both when in the enclosure and after release in the vicinity of Yalooni. Herds 1 and 2 had their own supplies in their herd ranges (Fig. 5.4). Henceforth, water consumption figures per oryx refer to an adult-equivalent (5.2.7).

Average water consumption as l/oryx per day was measured for herd 1 from May 1980, its first full month in the enclosure, to June 1986 (Fig. 6.11). The months with no value indicate periods when the herd was not drinking from the artificial supply. Excluding May to September 1982, consumption varied between 1.3 and 5.0 l/oryx per day. There was a strong seasonal effect, and consumption is correlated with the month's average maximum shade temperature (Fig. 6.12).

Consumption in the same month across years was fairly consistent, except in May to September 1982 (6.9.4). The low figure for November 1982 resulted from the herd entering a *bedu* camp on one night and drinking dry the camels' water. An approximate figure of 5.1 l/oryx was calculated. The figures derived for the herd overall cannot be used to see whether consumption declined with increasing exposure to the desert. The proportion of Oman-born oryx increased through the years and, even in the years 1980 and 1981, the herd contained a mixture of animals in their first and second summers in Oman.

These consumption figures are not physiological minimum requirements but the amounts drunk by oryx on an artificial diet, of an overall higher

Fig. 6.11. Herd 1's water consumption (litres/oryx per day) between May 1980 and June 1986. R indicates a month in which rain fell; * indicates a figure based on drinking for less than a whole month. In months with no value, oryx were not drinking from an artificial supply.

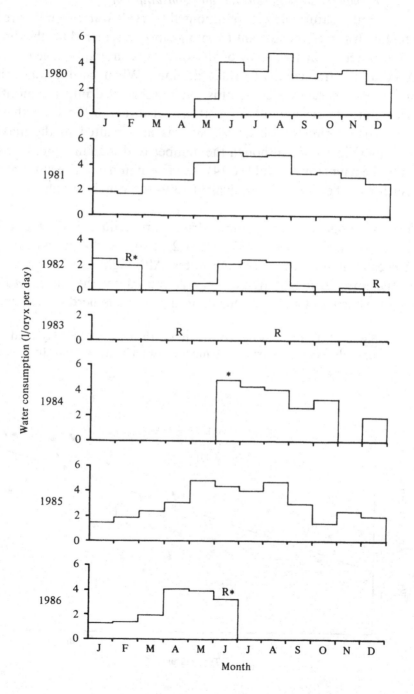

quality than most natural diets, with water within easy walking range. Periods of supplementary feed and water coincided because rainfalls allowed both artificial supplies to be discontinued.

6.9.3 *Flexibility in requirements and consumption*

Under natural desert conditions the oryx's water sources are liable to be highly dispersed and distant from one another. To reduce the distances walked to water or the frequency of visits, an oryx should be able to increase its intake as the frequency of drinks declines. When herd 1 was drinking outside the enclosure, over 26 months, the number of drinks per month was inversely related to the amount drunk at each visit (Fig. 6.13*a* and Table 6.14). However, the amount drunk on a visit was also related to the maximum temperature (Fig. 6.13*b*), although the number of drinks in a month was not related to the temperature (Table 6.14). The maximum average amount drunk per visit for the herd was 7 l, equivalent to 9–10% body weight of an adult oryx.

Drinking frequency and volume drunk are adjusted by the oryx to prevailing conditions (Table 6.15). Herd 2 probably drank less often than herd 1 because it grazed further from water. Although herd 2A was closer to water, the lower drinking frequency was caused by the male, Mustafan, reducing the chance of his group encountering the main herd at their common

Fig. 6.12. Herd 1's water consumption (litres/oryx per day) and monthly average maximum temperature (°C). ▲ = June, July, August 1982 (see the text).

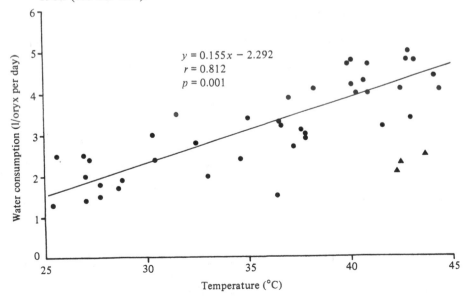

$y = 0.155x - 2.292$
$r = 0.812$
$p = 0.001$

Fig. 6.13. Herd 1's average water consumption (litres/oryx per drink) and (a) number of drinks per month or (b) average monthly maximum temperature (°C).

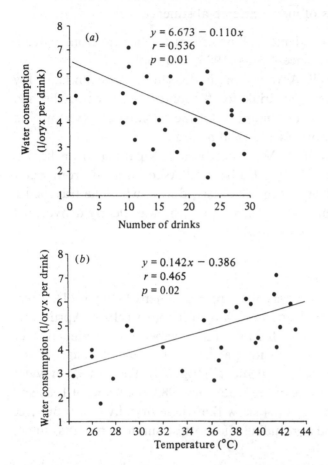

Table 6.14. *Partial correlation coefficients between number of drinks in a month (variable 1), average litres/oryx per drink (variable 2) and average monthly maximum shade temperature in °C (variable 3)*

| Variables correlated | Variable controlled for | Partial correlation coefficients | | |
		r	d.f.	p
1,2	3	0.477	23	0.02
2,3	1	0.461	23	0.05
1,3	2	0.042	23	n.s.

d.f., degrees of freedom; n.s., not significant.

water source (6.8). To obtain an average 2.1 l/oryx per day, the animals drank
10.9 l/visit, equivalent to 15% of body weight.

A few figures for single oryx drinking from small containers confirm greater
capacities under conditions of more extreme abstinence.

(1) In mid May 1984 Salama walked 45 km to Yalooni from range 8
    for her first drink since August 1983, and took 11 l.
(2) Hadya and her calf, Askoora, made the same journey in late May
    1984, also for her first drink for 9 months, since when she had
    calved and lactated for 6 months. Of their combined consumption
    of 11 l, Hadya probably drank almost all.
(3) Also in late May 1984, Mustafan became separated from herd 2
    initially through following Hadya and Askoora away from water
    (6.11). After 5 days lost in the desert, during which he travelled a
    minimum 140 km, much of it running, he was clearly dehydrated
    and drank 15.5 l from a bowl.

6.9.4   *Water independence*

Following rain, water was not supplied to herd 1 (Fig. 6.11). There
was standing water in the herd's current range for four weeks in April 1983,
but for only two weeks after the other rainfalls. The periods of independence
from artificial sources coincided with partial occupancy of range 1 and
occupancy of ranges 3, 4, 5, 6, 7, 8, 10 and 12 (Fig. 6.4). The periods of water
independence through the spring of 1982, autumn 1982, the whole of 1983 and
the spring and early summer of 1984 show that these oryx lived without free
water for prolonged periods, including the hottest summer months, under
suitable conditions.

Once standing rainwater has dried out, the oryx must balance a reduced
water intake with reduced loss in the face of changing requirements due to
ambient temperature. A number of factors influence their water intake.

Table 6.15. *Flexibility in water consumption and drinking behaviour in three*
herds in August 1985

Herd	No. drinks in month	Average intake (l/oryx per drink)	Average intake (l/oryx per day)	Average distance to water (km)
1	24	6.1	4.1	3.5
2	11	9.0	3.2	8.4
2A	6	10.9	2.1	6.1

(1) The moisture content of food. This is the main source of free water. Its contribution shows in the reduced quantities drunk by the oryx near Yalooni in May to August 1982 (Fig. 6.11), when they had green grazing from the February rain, unlike in these months of any other year.

Despite a maximum temperature of 43.9 °C in June 1983 and a last drink of rainwater 1.2 months previously, herd 1 did not need water from Yalooni, 25 km away (Table 6.16). The same was true in range 8, in April 1984, when the *Stipagrostis* contained 50% water, the air temperature was 39.9 °C and the herd had not drunk for 7.7 months. However, two months later when the temperature had risen and the moisture content of the same grass had declined to 35%, the herd returned to Yalooni. Range 8 was re-occupied in October 1984 as range 10, and the average maximum temperature had dropped to 31.2 °C in November when the grass moisture was again measured as 35%. These figures suggest that, without drinking, the oryx maintained water balance on a diet of 50%

Table 6.16. *The moisture contents of food and faeces of herd 1 in various locations and times, average monthly maximum shade temperature, with the periods since last rain and the oryx's last drink*

Date/ Range	Food/ % water	Temp. (°C)	Months since last		Faeces % water (±S.D.)
			Rain	Drink	
June 1983 6B	*Stipagrostis* 53 Herbs 75	43.9	2.5	1.2	48.5 (±4.74)
April 1984 8	*Stipagrostis* 50	39.9	8.1	7.7	40.0 (±4.27)
June 1984 8[a]	*Stipagrostis* 35	43.6	10.3	0[b]	—
August 1984 Yalooni	Grass hay 3 Alfalfa hay 17	42.5	12.5	0[b]	51.7 (±5.05)
November 1984 10	*Stipagrostis* 35	31.2	15.0	1.0	41.0 (±3.13)

[a] Samples were cut in range 8, 6 days after herd left it for Yalooni.
[b] Herd was at Yalooni with water supplied.

water at 40 °C or of 35% at 31 °C, but that balance could not be maintained under more extreme conditions such as 35% water at 42.5 °C. This assumes that water availability was the sole factor determining herd movements, whereas the protein content of the grass was certainly implicated (6.3.2), although these two quantities are rarely independent.

During dry periods, the Jiddat-al-Harasis has no obviously succulent plants which might be eaten for their water content, although the fleshy annual forb *Zygophyllum simplex* is taken, when abundant, 2–6 months after rain. Oman's oryx have not been seen digging for tubers, unlike the fringe-eared oryx (Root, 1972) and gemsbok (Williamson, 1987).

(2) Condensed water. Dews are common through the year in Oman's deserts, but the quantities of water available from condensed fog are much greater, although seasonal (Table 4.2). After thick fogs, all rock surfaces are wet and droplets hang from vegetation. On such mornings oryx are invariably found grazing, with wet muzzles, and it is possible that stones or vegetation are licked for their water.

(3) Timing of activity. The intense feeding activity as morning fogs lift, and the increased frequency of short-term separations on such mornings suggest that the arrival of fogs at night initiate grazing for the sake of their water. There may also be an additional water gain through the staple food, *Stipagrostis* sp., being hygroscopic: samples taken at 17:50 h on 9 April 1984 contained 47.1% (s.d. ± 3.50%) moisture, while three samples cut at 06:00 h the following morning, after a night with an overhead fog but with no condensed water on the grass, contained 52.4% (s.d. ± 1.17%) water.

The mechanisms employed by the oryx to reduce their water loss are probably many and diverse, as in other desert animals (MacFarland, 1964). Several observations show that re-introduced oryx respond to the exigencies of reduced water availability.

(1) Seasonal changes in activity patterns assist water conservation. The increase in time spent shading with temperature increase above 27 °C (Fig. 6.1) reduces the water requirement for evaporative cooling.

(2) Seasonal changes in coat density, colour and reflectivity (2.4.1)

promote heat absorption in winter but reduce it in summer. Coupled with the oryx's surmised thermo-lability (2.4.1) and the daily temperature variation even in summer, these adaptations and flexible shading times may result in a low water need for evaporative cooling.

(3) The frequency and duration of urination decrease when oryx are not drinking, and a more concentrated urine is likely. Faecal moisture content also varies (Table 6.16) from around 50% when in easy water balance to 40% when the balance is marginal.

6.10   Environmental perception and learning by oryx

6.10.1   *Navigation and site-recall*

The movements of oryx herds between ranges or on forays were clearly not random. Such moves and separations, whether accidental or deliberate, allowed some insight into the means and extent to which the re-introducing animals perceived, learned about and remembered their environment. Understanding these aspects helped in management decisions and showed the extent to which the oryx mastered their environment. The following conclusions are based on observations.

(1) Individual oryx remember every area that has been passed through or occupied once. No other explanation is possible for the precise return to former range areas after long intervals of time. Oryx return even to specific sites such as a heap of camel bones, which had last been chewed many months before. An oryx can walk straight to a water trough at night having travelled 45 km and been out of the area for many months. This precision in revisited areas may depend on the intimate knowledge of the area acquired when it was the current range (6.3.3).

(2) Routes between widely separated areas are remembered and used often. For example, herd 1's track on its return to range 3 from range 6 followed almost exactly that taken by the herd when leaving range 3 some 6 months before. Oryx may cue into old, wind-blown hoofprints on regular routes such as those into Yalooni from the south-west. This allows patrols to locate separated oryx efficiently by predicting where their tracks may be crossed if the motive for their separation has been correctly surmised.

(3) When on the move, oryx are constantly integrating information on their course and heading. When herd 1 left range 5 to return to the

familiar area of range 6 (Fig. 6.5*b,c*), its route passed 46 km through new country, but the herd returned direct to known territory where it then settled.

(4) Separated oryx have checked former ranges between 3 days and 24 months after last occupancy. Apart from watching for oryx, such animals appear to assess the freshness of tracks and droppings, for which the faecal pyramids of herd males offer useful clues.

(5) Young animals can navigate as accurately as adults into known areas, and experienced oryx will lead naive ones into new areas. This was proved by three oryx, all 11–12 months old, returning 45 km to herd 1 after release from 11 days in the pens for veterinary attention to one. One of the animals had been born at Yalooni to herd 2 and had never been more than 5 km away, but was prepared to follow his companions.

(6) Separated oryx either stay in an area known to the herd or search for oryx in collectively known areas. Even the margins of unfamiliar areas are not entered and searched, despite the lack of any boundary due to habitat change. A female and calf, separated near Yalooni when their herd-fellows, unnoticed, crossed a ridge into new ground, returned 4 km to the water trough to wait. Failing to meet or join the herd on several night visits to drink, the pair first spent a day touring the current range known to them. They then checked a *haylah*, much used in ranges 4 and 6. This was 25 km from Yalooni and the female had not been there for 24 months. All joined again on the pair's return.

These examples, in bare detail, show that the re-establishing oryx navigated accurately and remembered places, probably even to the scale of individual trees and patches of grazing, after limited experience and exposure. The ability to reach one destination by different routes shows that they were aware at all times of their relative position. They could assess whether an area was occupied by other oryx, and the method for locating their fellows was conservative but ultimately successful. Thus, unless their behaviour shows otherwise, single oryx or separated animals are fully in control of themselves, and will find other animals if left alone and allowed time. Such conclusions promoted non-interference in oryx movements by the ranger patrols.

6.11   The cautionary tale of Mustafan

Range development and exploration by the two herds, as described above, were gradual and methodical, and each major move was made in response to readily apparent ecological conditions or changes. In marked contrast, one incident involving a single oryx over 19 days demonstrated clearly the low chances of survival of an animal dispersing speedily while inadequately desert adapted and unable to navigate.

The incident commenced with the return to Yalooni of Hadya and her calf, Askoora, to drink from range 8, 45 km away, on 26 May 1984. At the time, herd 2 had been outside the enclosure for seven weeks, and Mustafan was mature but the second-ranking male of the herd. Herd 2 was 5–6 km from Hadya's direct track to the pens for water, but Mustafan, at least, detected the pair: on the morning of the 27 May he was accompanying Hadya and Askoora as they attempted to head back to herd 1 despite Mustafan's efforts to detain and herd them in any other direction. On the afternoon of day 2 the

Fig. 6.14. Route taken by Mustafan over 19 days away from herd 2 in Yalooni area. Broken lines are roads, as in Fig. 4.6.

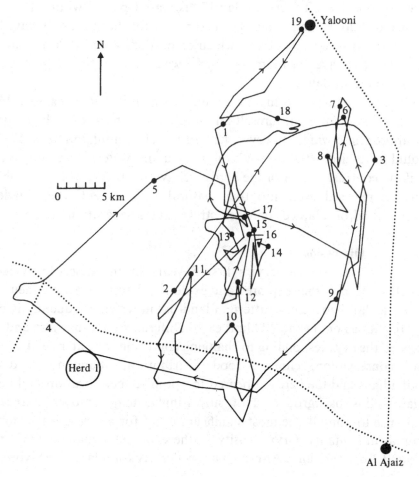

three were 33 km from Yalooni when Mustafan deserted the others, who were only 9 km from their herd, which they rejoined the following morning. Mustafan found himself in strange country and started off correctly for Yalooni, but when tracked up on day 3 he had veered away from Yalooni when only 10 km short. The sight of familiar patrol cars caused him to panic and gallop southwards. For 19 days in early summer he wandered (Fig. 6.14). On the afternoon of day 3 he entered herd 1, which had been directed towards him in an attempt to calm him, but, although he met with no aggression, the herd soon paid him no attention and he turned north.

On the afternoon of day 5 he was given water (6.9.3), and daily tracking continued. His pattern was to move north by day, during which attempts to herd him this way were made, followed by moves south by night into the sea-breeze. On day 7 he was given alfalfa because of his loss of condition, which with the daily presence of a patrol calmed him. His movements steadied on day 8 in the area through which he had passed with Hadya and Askoora a week before. He left there on day 17, as herd 1 passed within 3 km on their return to Yalooni from range 8. The stimulus for the move may have been the start of a north-west wind into which he headed. When 10 km from Yalooni the area may have been familiar for he was found in the range of herd 2 on the morning of day 19.

Once his panic at being in an unknown area had subsided, Mustafan appeared able to orient himself and find his way back with the assistance of an appropriate wind. However, he had travelled a minimum of 440 km and would have died without provision of food and water. His example proved a striking contrast to that of the herds' behaviour. It revealed the problems for animals released even into their natural environment when inadequately acclimatised or adapted, or when stress causes separation and panic.

6.12   Discussion

The released oryx's first 5 years in the desert provided many insights into how they explored and exploited their new environment. Herd 1, in particular, had a clear pattern of land-use in which it sequentially occupied discrete areas of around 300 km² each. Differences in range size and location showed the oryx responding to changing conditions of water need, availability and conservation, climate, food distribution, abundance and quality. Individuals and the herds appeared to optimise their location and behaviour against this background. Not surprisingly, major moves were made in response to rainfall, the most significant event for all the desert's human and animal inhabitants. Oryx density in these ranges remained fairly constant around 1 oryx/22 km². Although a low density for a large herbivore, it is the same as that calculated for the scimitar-horned oryx in Chad (Table 2.1). Even though the Jiddat-al-Harasis' productivity is enhanced by its peculiar climate (4.4), its sparse resources oblige the Arabian oryx to live at low densities. It will be interesting to see if, as numbers increase, density increases or the occupied area expands in proportion.

After $4\frac{1}{2}$ years in the desert, herd 1 knew almost 2000 km^2 and this increased further with rain in early–mid 1987. The method used to measure range areas yields underestimates because it excludes the areas passed through on forays and journeys or those visited by subgroups. Because some oryx separate more often (6.6.2) and subgroups became common from early 1986, individual oryx will increasingly develop their own ranges and areas of particular familiarity. No oryx showed signs of having settled by breeding age in a home-range for the rest of its life, as in many large mammals. The desert's capricious temporal and spatial patterns of water and grazing prevent this option. It looks, rather, as if the oryx can potentially increase their home-range throughout life as conditions demand. As an 'explorer' species in an extreme and unpredictable environment, it resembles moose, which range widely in search of their short-lived successional habitats (Geist, 1971). The forays which herd 1 undertook usually after moving into a new range were exploratory to assess habitat conditions in neighbouring areas. A lifetime spent occupying new areas in this way will result in an enormous home-range, and one which is used differently from the large ranges of mass migratory species of more predictable environments, such as the Serengeti wildebeest (Maddock, 1979).

The infrequency of separations from the herds, for whatever reason, show they were very cohesive units. Cohesion was aided by weather conditions determining and synchronising individual behaviour. In addition, the periods spent in the enclosure forged strong bonds between herd members, which were the basis for the marked intolerance between herds and the low rates of exchanges between them. The structured pattern of a grazing herd showed that the assembled groups contained the right mix of sex and age for stability over the first, critical years in the desert. The 'integrated' herd was achieved and served its purpose (3.7.3).

Subgroups evolved in both herds late in a long dry period. There seemed good reasons for this at the time (6.7), and as all the animals were then being provisioned where they were each day, each subgroup learned that it could be self-sufficient when away from the herd. After the rain of June 1986, when both herds moved far south of Yalooni (Figs. 6.4 and 6.8), these subgroups became the norm. Moreover, many contained equal numbers of original herd 1 and 2 members. Monitoring became more difficult, and two subadult males were not seen for 6 weeks, re-appearing close to where they were last seen. There was no sign of any distress from being on their own. They and other small groups were evidently learning that they could feed for longer than a large herd on small areas of good grazing, which were then common. This may prove a milestone development in the oryx' progressive independence, for in March 1987 heavy rain drew all the Harasis 200 km north-east to the Wadi Halfayn. The oryx were left with the Jiddah, which had received only light and very scattered rain, and they continued to exploit the resulting light grazing with small groups which met and merged and split in new combinations (T. H. Tear, personal communications 1987). This behaviour suggests that the oryx's recent experience overcame the original, strong bonds between herd-members in favour of new and more adaptive ways to exploit the desert. It is possible, therefore, that the re-introduction methods overemphasised herd hierarchies and cohesion, but this approach un-doubtedly served the re-introduction well (1.3.2).

Both captive-bred and wild-born oryx had to learn to be on their own in the desert. Many observations showed them learning to use their environment and its clues, as when detecting and following fresh tracks. All learnt to move to rainfall areas. This habit must be strongly re-inforcing and may lead to oryx moving towards black clouds before they smell rain or fresh grazing (6.3.2). Return forays from the east, south and south-west edges of the Jiddah show their recognition of ecological boundaries and confirm the suitability of the Jiddah as the re-introduction site. The oryx's feats of navigation and site recall show that an animal is always aware of its location, with a long-term memory of areas after single exposure to them. There has been no record of any oryx failing to navigate straight into Yalooni from afar, and herd 1's return after 22 months away was evidence of the long-term fidelity to the release area and its value as a management method (1.3.5).

In the desert the oryx progressively exploited their physical and physiological potentials. They learned to use their colour and conspicuousness to locate one another, and their walking stamina for long journeys between ranges or back to Yalooni for water. The released oryx showed how they can be water-independent for long periods by modifying their behaviour and through careful habitat selection. The occasional returns to Yalooni by heavily pregnant or lactating females showed they, at least, knew water was available. The herd as a whole weighed up the benefits from walking in to drink against its energetic cost, with an independence due to a flexible water metabolism. The oryx born in Oman may yet show greater water-independence than those imported, but clearly the re-introduced oryx survived for long periods without water in the conventional, anthropocentric sense. They exploited the unusual sources of small amounts of moisture in the Jiddah and their own water conservation mechanisms.

This chapter has shown how the re-introduced oryx established themselves and exploited their ancestral habitat in a rational and masterful fashion. After 5 years in the desert, their behaviour is still developing as they experience new conditions, test themselves further, and their perception of the desert evolves. This, however, is not the whole story, for the success of the herds depended on the performance of their component animals. The cost of this success in terms of individual survival and productivity, and the selective forces that were faced, are described in the next chapter.

7

Individual oryx in their natural environment

7.1   Introduction

While the last chapter described the movements of oryx herds and their response and use of their environment, it more or less ignored the occurrence of births and deaths. This chapter is concerned with individual performance and its determinants in the re-establishing oryx. It also attempts to assess the impact on the immigrant oryx of being transported from 33° N to 20° N, and some comparative data from the captive populations of the USA are used.

Accurate identification of causes of death in a re-establishing population is important, and was possible in Oman because the oryx were monitored closely. Analysis of the causes of mortality show whether there is any pattern, which management might overcome, and the nature of the problems faced by the population during establishment. The project has run long enough for these same data to provide mortality rates for comparison with those of captive oryx.

Oman's immigrant oryx resulted from several generations of captive-breeding during which they and their ancestors were exposed to novel selective forces, especially those favouring success in captivity (Frankham *et al.*, 1986). Such exposure could affect adversely the adaptability of captive-bred oryx to the desert, and hence jeopardise the re-introduction. Here, individual performance is assessed in relation to the genetic composition of the population and the level of inbreeding. Differential mortality might narrow the genetic base or increase inbreeding level, with potentially lethal consequences for the re-introduction several generations later. Finding a significant impact of inbreeding on performance would encourage manipulation of the breeding system or influence the selection of further immigrants.

In theory, such a result from the wild population should feed back to modify genetic management of the captive population.

7.2 Reproduction in Oman

7.2.1 *Oryx numbers*

The number of oryx alive in Oman at the end of each year was the balance between the numbers of immigrants and animals born in Oman less those dying in either class. Details of the immigrants are given in 5.2.3. For the present purpose, the two immigrants which were unfit for incorporation into a herd for release (5.2.5) were not counted as having come to Oman. At the end of 1980, the 10 oryx comprised nine surviving founders and one born in Oman, while at the end of 1985 and of 1986, the herd had increased to 31 (Fig. 7.1). The proportion of Oman-born animals was 48 % in 1983 and 71 % in 1986. (For details of all oryx, both immigrant and Oman-born, see Appendix 3.)

The population increased exponentially at 22 % per year, with a doubling

Fig. 7.1. Total number of oryx alive in Oman at end of each year, excluding two oryx unfit for release; shaded portion is numbers born in Oman; figures in parentheses are numbers of immigrants received in each year and fit for release.

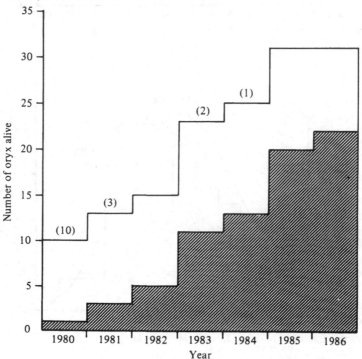

time of 3.45 years. On this basis, Oman's herd was increasingly slightly faster than the World Herd between 1964 and 1978 (3.5.3). A demographic model used later for modelling population growth shows a lower rate of increase but was based on life table values calculated for both the oryx in Oman and their parents (9.3.2).

Fig. 7.2 *a*, *b*. For caption see opposite.

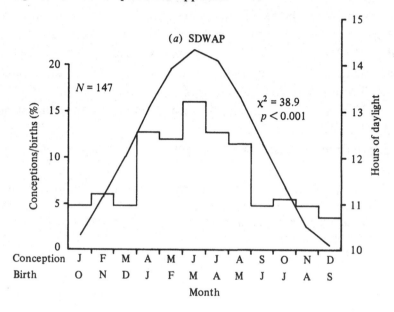

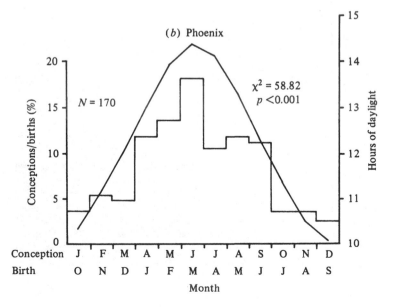

7.2.2   *Gestation length*

In Phoenix Zoo, the Arabian oryx had an oestrous cycle of 25–32 days (Turkowski & Mohney, 1971). In Oman, the dates of all observed matings or mountings were recorded as indicators of a potential conception, and those births for which a previous mating must have caused conception could be identified. Precise observations of 15 intervals between mating and calving yielded an average gestation of 264 days, with a range of between 255 and 273 days. This mean is slightly higher than two records of 255 and 260 days from Phoenix Zoo (Turkowski & Mohney, 1971), but is used in subsequent calculations as 8.7 months or rounded to 9 months.

7.2.3   *Seasonality of calving*

For the first years in Oman, any season of births might have reflected residual patterns from the USA, rather than responses to local conditions. While SDWAP and Phoenix both lie at latitude 33° N and have the same daylength and similar minimum winter temperatures, Phoenix has a lower rainfall and hotter summer (Fig. 5.1). Yalooni at 20° N has a climate similar to that of Phoenix but is more arid with 70% less rainfall. The Oman summers are hotter and the winters milder.

Births in both US herds were unevenly distributed through the year (Fig. 7.2). Although some oryx were born in each month, a distinct peak accounted

Fig. 7.2. Monthly frequencies of conceptions and births, with daylight length at (*a*) San Diego Wild Animal Park (*b*) Phoenix Zoo (*c*) Yalooni. χ^2 values test for monthly differences in conception/birth rates; the sample size for Yalooni was too small.

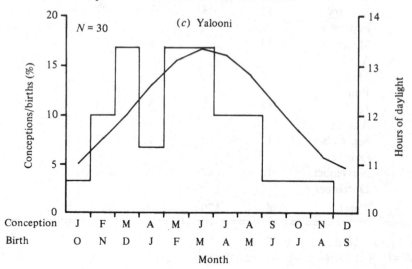

for 65% births at SDWAP between April and May, and 66% at Phoenix, although the peak extended here through September. From the arrival of the first oryx in Oman in March 1980 to the end of July 1986, 29 calves were born alive, with a further three stillbirths. The first two calves born in Oman were conceived in the USA and are excluded from this analysis. The sample of 30 births is too small to test for differences between months (Fig. 7.2). Although more calves were born at Yalooni in November–December, the birth peak in March and trough in September were common to Oman and both US herds. The Yalooni distribution of births is not different from those of SDWAP and Phoenix (Table 7.1).

When birth frequencies are offset by 9 months to obtain conception dates (Fig. 7.2), conceptions peaked in the US herds when daylength exceeded 13 h, although at Phoenix a high conception rate continued in September with 12.3 h of daylight. Although daylength and conception coincided less well at Yalooni, oryx at all three sites appeared to be long-day breeders (cf. Amoroso & Marshall, 1960).

One adaptive value to the oryx of conceiving in long-day months might be to avoid calving in hot months in Oman (Fig. 7.3a). The Phoenix data do not show such a relationship (Fig. 7.3c). The same phenomenon at SDWAP may be spurious because of the influence of 5 months with high conception rates and

Table 7.1. *Contingency table of number of births in SDWAP, Phoenix and Oman, combined for 2-month periods for sample size*

Months	SDWAP	Phoenix	Oman	Total
January–February	37	43	7	87
March–April	43	49	8	100
May–June	24	39	4	67
July–August	15	12	2	29
September–October	12	10	1	23
November–December	16	17	8	41
Total	147	170	30	347

$\chi^2 = 10.533$, 10 d.f., $P > 0.05$.
SDWAP, San Diego Wild Animal Park.

7 months with low rates, producing uneven variances around the regression line (Fig. 7.3b). In any event, captivity has an ameliorating influence on climate and nutrition, reducing the hazards of summer calving. In Oman, this calving pattern would reduce the liability of young calves to heat stress and would benefit the mothers' water economies by lessening lactational demand

Fig. 7.3. Percentage of births and average monthly maximum air temperatures at (a) Yalooni, (b) San Diego Wild Animal Park, (c) Phoenix Zoo. n.s., not significant.

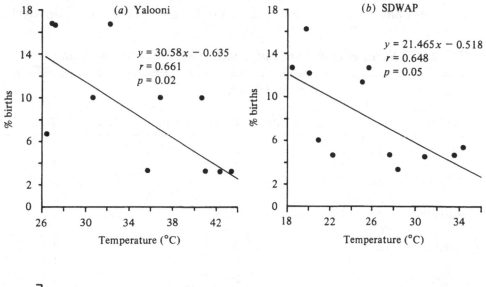

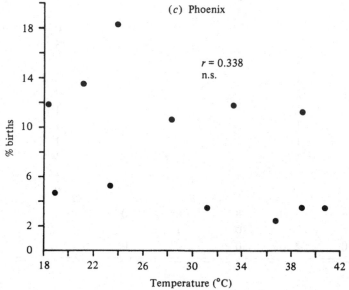

in the hottest season. Improved winter grazing from either rain or fog moisture (4.4) would benefit the nutritional status of females during lactation.

7.2.4   *Calf sex ratios*

The sex of two calves born in Oman was not known. One calf was premature, presumed stillborn and never found, while the other was separated from its mother at 4 days and disappeared before it could be sexed. Two calves were sired by two males in the USA, and three males sired the other 28 calves in Oman (Table 7.2). Of these, 18 were male and 12 female. The combined offspring of Jadib and Lubtar totalled nine males and 11 females, while Hamid sired eight males and no females. This last ratio deviates

Table 7.2. *Sexes of calves born in Oman to each mother and father*

Mother	Calf sex	No of calves to each father					Total
		Jadib	Lubtar	Hamid	Antar	Museba	
Salama	m	2		3			5
	f	1			1		2
Talama	m	2					2
	f						0
Rahaima	m	1					1
	f					1	1
Hadya	m	1		2			3
	f	1					1
Alaga	m			2			2
	f	1					1
Farida	m						0
	f		3				3
	?		2				2
Selma	m		1				1
	f	1	2				3
Zeena	m		1				1
	f		1				1
Kateeba	m		1				1
	f		1				1
Mundassa	m			1			1
	f						0
Total	m	6	3	8	0	0	
	f	4	7	0	1	1	
	?	0	2	0	0	0	

m, male; f, female.

significantly from equality (binomial test, $p = 0.004$). In addition to these eight males, the females Rahaima and Talama were carrying male foetuses by Hamid when they died.

7.2.5 Age at first calf and inter-calf interval

All breeding females in Oman fell into three groups, depending on their origin and history (Table 7.3). Group A females, numbering three, either had their first calves in the USA or travelled to Oman pregnant. On average, they calved for the first time at 25.0 months. In contrast, the three females of group B, which arrived in Oman at ages between 11.6 and 17.9 months, did not produce their first calf until 41.9($\pm$3.8) months, clearly later than group A. A fourth female in this group, Hababa, aborted her first foetus at 6.7 months. She would have been 36.8 months at predicted full term, but had not

Table 7.3. *Age at first calf and inter-calf intervals of female oryx divided into three groups*

Name	Age at first calf (months)	Interval between successive calves (months)					
		1–2	2–3	3–4	4–5	5–6	6–7
A. Born in USA and calving there or travelling pregnant to Oman							
Farida	25.8	10.8[a]	15.7	10.0	9.9	10.1	14.8
Salama	22.8	21.9	9.2	10.7	8.9	9.0	11.4
Rahaima	26.3	29.2					
	av. 25.0						
	s.d. 1.54 ($n = 3$)						
B. Born in USA, first conceiving and calving in Oman							
Talama	46.2	10.4					
Hadya	42.6	8.4	12.7	12.9			
Zeena	37.0	13.1					
Hababa	(36.8)[b]						
	av. 41.9						
	s.d. $\pm$3.79 ($n = 3$)						
C. Born in Oman							
Selma	32.7	8.8	9.0	9.1			
Alaga	30.9	13.2	13.8				
Kateeba	27.4	9.2					
Mundassa	33.8						
	av. 31.2						
	s.d. $\pm$2.43 ($n = 4$)						

[a] Farida's first and second calves were born in the USA.
[b] Age on estimated birthdate if Hababa had not aborted at 6.7 months.

calved subsequently at 51.2 months old. Mustarayha, also in this group, was never released from the pens (5.2.5), and ran with an infertile male for many months. She subsequently conceived and bred by a different male of Bahrain stock in the Oman Endangered Species Breeding Centre in Muscat.

Group C comprised the females conceived, born and growing up in Oman. For four such breeding oryx, their average age at first calf was 31.2 ($\pm$2.4) months. This was later than the group A females, but earlier than those of group B. In addition to these four, another two young females had not produced their first calves at 39.8 and 31.2 months. These limited data suggest that oryx born in Oman had their first calves later than in a zoo in the USA, but that females who travelled before their first successful mating had severely

Fig. 7.4. Monthly births in San Diego Wild Animal Park.
(*a*) Distribution preceding an inter-calf interval of $\geq$ 10.0 months,
(*b*) hypothetical distribution if the next calf had been born after
9.0 months, (*c*) actual distribution of births of next calf after
intervals $\geq$ 10.0 months.

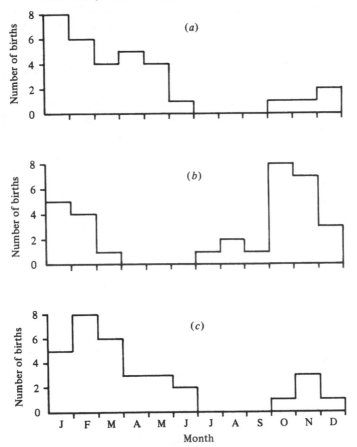

delayed first calvings. The possible causes of this disadvantage are discussed below (7.2.6).

In the early days of the World Herd, the interval between calves in Phoenix Zoo averaged 14.1 ($\pm$2.3) months because of the deliberate separation of females from males before and after calving (Dolan, 1976). Inter-calf periods were shorter at SDWAP, with 64% less than 10.0 months (Table 3.2). The periods longer than 10.0 months provide additional evidence for the overriding importance of daylength (7.2.3). Thirty-two births followed after intervals greater than 10.0 months, since the previous calf to the same female (Table 3.2). If these calves had followed a 9 month interval, only 31% would have fallen in the five 'high-birth' months, compared to 84% of the previous calves to the same females (Figs. 7.4*a*, *b*). In contrast, the actual intervals in excess of 10 months restored this frequency to 78% (Fig. 7.4*c*). Thus, in SDWAP's captive population the inter-calf interval was determined primarily by daylength, possibly to the total exclusion of other factors such as nutrition or body condition.

The group A females in Oman showed the most variation in inter-calf interval (Table 7.3). Farida had 10.8 months between her first and second calves in the USA. She was then either not mated or failed to conceive during the period of preparation for travel to Oman, but was mated soon after arrival in Oman, causing the longer interval of 15.7 months. Once in Oman, her next three intervals returned to 10 months. In contrast, Salama and Rahaima, who both arrived in Oman pregnant with their first calves, did not produce their second until 21.9 and 29.2 months, respectively, after the first. Rahaima died after calving only twice, but Salama's intervals after her second calf were consistently between 8.9 and 11.4 months. This delay in first conception in Oman by these two immigrant females was similar to the delay in breeding of the group 2 females, and is also discussed below.

Apart from these figures for the group A females, most intervals lay between 9 and 14 months. The lower figure resulted from successful *post-partum* mating. There were indications of differences between females, such as Alaga, whose calves were consistently more spaced than Selma's. Inter-calf intervals were the same between females of the three groups, indicating that when a female had conceived and bred once in Oman, her rate of calf production was independent of her origin or past. Moreover, the shortest intervals recorded in Oman were equal to those under the best captive conditions in the USA.

7.2.6   *Immigrant females' first calves in Oman*
          The data above showed that, with the exception of one female, Farida, the first conception in Oman which led to a calf was delayed in immigrant females. This occurred whether or not the female arrived pregnant from the USA. Was this effect due to some factor of age of the animals, some factor associated with the journey itself, or social or environmental conditions in Oman?

Effect of travel
Of the group A females, Farida travelled after having two calves in the USA. Rahaima and Salama travelled pregnant with their first calves which were born 5.7 and 2.2 months after arrival, respectively. Both calves were abandoned by their mothers and had to be hand-reared. Although first young are more prone to being abandoned in domestic sheep (Winfield, 1970), only two calves were abandoned subsequently in Oman, and both proved not viable (Bin Talama 1 and Bin Alaga, see 7.3.2). This suggests that mothering ability in these primiparous females was adversely affected by travel.

Table 7.4. *Factors potentially influencing time of first conception in Oman in two groups of immigrant female oryx*

Name	Period in Oman to birth of first calf conceived there (months)	Period in captivity in Oman (months)	Age on arrival in Oman (months)	Age at first successful conception (months)	Period from release to first successful conception (months)
A. Born in USA and calving there or travelling pregnant to Oman					
Farida	8.7	30.8	43.5	17.1	—
Salama	21.9[a]	22.8	20.6	14.1	—
Rahaima	29.2[a]	13.8	20.6	17.6	12.5
B. Born in USA, first conceiving and calving in Oman					
Talama	34.6	22.8	11.6	37.5	3.1
Hadya	27.2	13.7	15.4	33.9	4.8
Zeena	19.0	8.6	17.9	28.3	1.7
Hababa	22.1[b]	8.6	14.7	28.1	(4.7)

[a] Counting from birthdate of calf carried from USA.
[b] Hababa's figures based on predicted calving-date in absence of abortion at 6.7 months.

Effect of age at travel

Calculation of the ages of the females of groups A and B on their arrival in Oman and ages at conception of their first calf (Table 7.4) shows that all three of group A conceived in the USA at between 14 and 18 months old. The four females of group B travelled to Oman at ages between 12 and 18 months, but then did not conceive for the first time until 28 to 38 months old. This suggests that the journey, or conditions before or after it, at ages when these females might have conceived for the first time in the USA, prevented successful conception in Oman.

Effect of captivity in Oman

The periods spent in captivity in Oman by the immigrant females varied between 8.6 and 30.8 months, but were not related to the times between arrival in Oman and birth of the first calf conceived there (Table 7.4, $r = -0.311$, 5 d.f., not significant). Thus, the period in captivity was not responsible for the delay in first conception in Oman and, indeed, Farida, who spent the longest time in the pens and enclosure, had three calves there.

Both Salama and Farida were pregnant when released from the enclosure into the desert. Four of the other five females of groups A and B (Table 7.4) conceived between 1.7 and 4.8 months after their releases. Rahaima conceived successfully only after 12.5 months in the desert. Although this suggests that females were not conceiving until outside the enclosure, the enclosure clearly provided adequate conditions for Farida for she gave birth to her third, fourth and fifth calves there, with short intervals between them (Table 7.3). Moreover, the Oman-born Selma, who was released with herd 1 but was then confined again to join herd 2 (5.3.2), gave birth to her first and second calves in the enclosure, and the second and third were conceived there, with inter-calf intervals of only 8.8 and 9.0 months.

7.2.7   *Occurrence of oestrus*

The possibility that the delay in the first calf to immigrant females was due to oestrus failure can be investigated from records of sexual behaviour while the oryx were under intensive observation in the enclosure and after the herds' releases. Table 7.5 compares the frequency of oestrus of Farida, who bred successfully in captivity, with the performance of six other females. First conceptions leading to births in the latter occurred between 10.4 and 26.1 months after their arrival in Oman, compared to 0.3 months for Farida. For the six females, there were between three and nine observed oestrous periods per female, with intervals of 2.3 to 4.5 months between successive oestrus. The longest periods with no observed sexual activity for each female varied between 4.9 and 9.7 months. With an oestrous cycle of 25–32 days in Phoenix Zoo (Turkowski & Mohney, 1971), these figures are evidence that oestrus was suppressed after arrival in Oman. In addition, females were mated when in oestrus but were failing to conceive. The dominant males of both herds sired at least one calf each while in the enclosure and then sired calves by the delayed-breeding immigrants after release into the desert. The failure to conceive and the low frequency of oestrus were probably associated, and unlikely to be due to male inadequacy at the same time.

7.2.8   *Causes of reproductive delay*

The delay in first breeding in the immigrant females might have been related to adjustments to the Oman daylength. However, there was no evidence of an altered calving season (7.2.3), and daylength in no month differed by more than one hour between Yalooni and SDWAP (Fig. 7.2). If a threshold effect operated, there might be some significance in daylength in California exceeding 13 h in 5 months, but in only 3 months a year in Oman. The photoperiod may have influenced the oryx in other ways, for it was obvious that the second group of immigrants (Table 5.1), which reached Oman in December 1980, never attained their full summer coat in 1981. The two groups arriving in July and September developed typical sparse coats in the next summers.

Table 7.5. *Occurrence of oestrus in immigrant female oryx*

Name	Period in captivity in Oman (months)	Period from arrival to first successful conception (months)	No. oestrus seen up to and including first successful conception (months)	Av. period between oestrous periods (months)	Max. period between oestrous periods (months)
Salama	22.8	13.3[a]	3	3.2	9.7
Rahaima	13.8	20.6[a]	9	2.3	7.6
Talama	22.8	26.1	9	2.9	8.6
Hadya	13.7	18.6	7	2.7	5.9
Zeena	8.6	10.4	4	2.6	4.9
Hababa	8.6	13.6[b]	3	4.5	9.7
Farida	30.8	0.3	1	—	—

[a] Counting from birthdate of calf carried from USA.
[b] Counting to conception of calf aborted at 6.7 months.

The low levels of aggression and rapid development of stable dominance hierarchies amongst the members of the two herds in the enclosure indicated no social factors such as stress in the suppression of oestrus.

On travelling from the USA to Oman, the oryx's diet in captivity changed. The SDWAP diet comprised a mixture of a dry oat hay with 8–10% crude protein and lucerne pellets of minimum 13% crude protein. The hay had to be severely rationed before the oryx ate the pellets (L. M. Nelson, personal communication 1980), and was offered *ad libitum*. The Oman diet consisted of a mixture of hay and lucerne with an overall 14.5% crude protein (Table 5.3) fed in amounts to maintain condition. Thus, the diets in the two countries were unlikely to differ in their main nutrient levels.

The quality of the oryx diet varied in the enclosure with the availability and quality of its grazing, and the level of supplementation. As both herds were released when natural grazing was poor, the change in diet over the transition from captivity to the wild was not great. Faecal nitrogen level indicates the quality of the diet of grazing fringe-eared oryx (Stanley Price, 1985*a*, *b*), and this was measured on various occasions for both herds in Oman before and after release (Table 7.6). These figures show that diet quality was no worse in the enclosure than in the desert.

Table 7.6. *Faecal nitrogen concentrations in herds 1 and 2 before and after release, with various diets*

Herd	Location	Diet	Period	Nitrogen % Av.	s.d.
1	Enclosure	Hay, lucerne, almost no grazing	3 mo. before release	1.24	±0.104
1	Desert	Grazing only, 1 mo. after rain	2 mo. after release	1.13	±0.202
1	Desert	Grazing only, 2 mo. after rain	3 mo. after release	1.20	±0.162
1	Desert	Grazing only, 5 mo. after rain	6 mo. after release	2.00	±0.378
2	Enclosure	Hay, lucerne, almost no grazing	1 week before release	1.13	±0.064
2	Desert	Grazing and hay, lucerne; no rain for 27 mo.	1 mo. after release	0.88	±0.085

mo., months.

Even if the causes remain obscure, most immigrants suffered breeding delays in Oman. The performance of the sole female who had bred in the USA suggests that the effect of a new regime may have been age-related or had more impact on females who changed regimes around the age of sexual maturity. This setback should clearly be avoided in future immigrants, and would be most easily achieved by selecting suitably aged females.

7.3   Mortality

7.3.1   *Death rates*

Of the 18 oryx coming to Oman from USA or Jordan, 8 (44%) had died by 31 July 1986. The periods alive in Oman varied widely: Malak died in the pens only one month after arriving, while Hadya spent 66 months in Oman, for 53 of which she was in the desert. So far, immigrant mortality has affected both sexes equally: nine of each sex came to Oman, and four males and four females have died subsequently. Ages at death varied between 10.5 and 170.6 months (Table 7.7), although in the latter instance the animal was never released from the pens (5.2.5) and died in the Oman Endangered Species Breeding Centre in Muscat.

Twenty-nine live births, three stillbirths and one abortion at 6.7 months were recorded in the same period. Of the 29 live births, 22 were still alive, with the oldest 43 months old at 31 July 1986. Of the seven Oman-born oryx which died, five (71%) died before reaching one month in age. The other two died at 1.7 and 29.1 months (Table 7.7). This suggests that survival in Oman-born oryx above 30 days old was high.

Bovidae at SDWAP suffered a perinatal mortality (the sum of neonatal mortality to 10 days after birth, stillbirths and abortions) of 19% (Benirschke *et al.*, 1980). The comparable figure for the oryx in Oman was 21%. Eight out of 32 (25%) oryx born in Oman were stillborn or had died at less than 30 days old. Mortality to this age in scimitar-horned oryx at Whipsnade Park, UK, under semi-free range conditions with grazing, was 13.3% for a sample of 15 births, but for all artiodactyls was 26.9% (Kirkwood, Gaskin & Markham, 1987). So, although the causes of death in captivity and the desert may be very different, early survivorship in Oman was certainly no worse.

Including the stillbirths, seven of the 23 calves (30%) born to immigrant mothers died. The figure for calves of Oman-born mothers was three out of nine calves (33%). The samples sizes are small but suggest that female origin did not influence calf survival. The average age of the surviving calves born to Oman-born females is currently lower, but as most calf mortality occurs within 30 days of birth, this results in little bias.

Table 7.7. *Oryx mortality in Oman and causes of death, distinguishing between immigrant oryx and those born in Oman*

Name	Sex	Age at death (months)	Cause of death/comment
A. Oryx born outside Oman			
Malak	m	10.5	Snake-bite while in pens. Blood positive for *Echis carinatus* venom
Sajba	f	31.3	Loss of condition following damage to scapula–thorax attachments while in captivity 12 months before death. Acute *Klebsiella pneumonia* infection and septicaemia. Died in enclosure
Museba	m	56.0	*Actinobacillus* infection, causing 'wooden tongue'
Rahaima	f	66.2	Myopathy, but identical symptoms to Talama suggest *Clostridium botulinum* toxaemia
Talama	f	69.4	*C. botulinum* toxin (type D) found in rumen and intestinal contents, serum
Nafis	m	89.2	Loss of condition, anaemic, broken rib with large pleural adhesions
Kadil	m	170.6	'Old age', infected prepuce, lung tumour, advanced liver degeneration; never released from pens
Hadya	f	82.1	Not certain: *Salmonella* positive. *C. botulinum* present in rumen contents but no toxin detected. Stressed through fright
B. Oryx born in Oman			
Haleema	f	29.1	Unspecific epilepsy or central nervous system disorder
Bin Talama 1	m	0.1	Obstructed intestine
Bin Salama	m	0.9	Peritonitis and pneumonia following horn wounds, sepsis and fly-strike
Bin Talama 2	m	—	Stillborn
Bint Alaga	f	—	Stillborn
Bin Farida 1	?	—	Stillborn
Kasheeda	f	0.7	Severe dehydration following 72 h separation from mother, then *Salmonella* infection and diarrhoea
Bin Farida 2	?	0.1	Separated from mother, never found
Bin Alaga	m	0.2	Weakling, abandoned by mother, hand-rearing failed. Enteritis and dehydration on autopsy
Saleem	m	1.7	Small for age, death due to attack by ravens, *Corvus ruficollis*. No bacterial infection

m, male; f, female.

7.3.2   *Causes of death*

Causes of death in both immigrant and Oman-born oryx were carefully recorded (Table 7.7). The causes were diverse, but there was no sign of epidemic disease, no losses attributable to parasite burden, only one calf loss due to predation and no instance of adult animals being lost and/or starving in the desert.

Of the eight dead immigrants, the death of Malak due to snake-bite was accidental, and Kadil's death was ultimately due to old age. Physical injury was the primary cause of death in the case of Museba, who suffered a pharyngeal puncture probably while feeding, of Sajba, who damaged her shoulder 2 days after arrival in the pens, and of Nafis, who probably received a blow to his rib-cage during an interaction with a dominant male (but see 7.4.2). Subsequent bacterial infections killed Museba and Sajba. Hadya was carrying *Salmonella* sp., at the time of death, despite being in excellent condition. Talama's cause of death proved to be *Clostridium botulinum* toxaemia, and the similarity of Rahaima's case-history suggests that she also died from this. Both females were noted scavengers of rubbish, such as strips of cloth or gazelle carcasses. The latter are ideal sources of the bacterium. Thus, bacterial infections were serious primary or secondary causes of mortality.

The causes of death amongst the younger, Oman-born animals were different. No specific reason for Haleema's progressive disorder over its 5 month course was discovered from the preserved autopsy material available. Bin Talama 1 was abandoned but then died during hand-rearing, due to an obstructed intestine. The abandonment of Bin Alaga by his mother, who had already reared one calf, suggested that this calf was not viable, and it was small at birth. The death of Bin Salama was by infanticide, following a change in dominant herd male (6.8).

The factors causing the separation of Kasheeda and Bin Farida 2 from their mothers are discussed later (8.5). Although Kasheeda was successfully reunited with her mother, the 72 h absence, during which she travelled at least 33 km, caused severe dehydration, and the associated stress brought out *Salmonella*. Saleem, who was killed by ravens at 1.7 months frequently trailed behind his mother or the herd when on the move, and the *post-partum* separation of the pair from the herd was more prolonged than usual, as if the calf was less hardy. In other cases of corvids apparently predating or scavenging young ungulates, the birds took animals that were either already dead or doomed to die anyway through exhaustion of fat reserves and starvation (Houston, 1977).

The three stillbirths occurred in a small time-span and shared some aspects with the premature birth at the same time of a fourth calf which survived (Table 7.8). All four females were mated under conditions of new grazing following heavy and widespread rain on 1 April 1983. Two other herd 1 females who were mated 6 weeks before this rain produced normal healthy calves after full gestations. In the autumn of 1983 many local goats, mated after the same rain, showed vaginal bleeding and were producing stillborn or inviable kids. The Harasis blamed consumption of the herb *Lotus garcinnii* soon after conception. This plant was abundant in herd 1's range after the April rain, and it was likely the oryx ate it at least on first exposure. It was also recorded at the time in the enclosure where Farida was the only adult female of herd 2. The coincidence of premature calving in four oryx in widely separated places suggests the Harasis explanation has some merit, even if a different ephemeral forb might have been to blame. The fact of Farida's premature calving in the enclosure, where there was no scope for interaction with domestic stock, reduced the possibility of contagious abortion, *Brucella* sp., as the cause in the other females. No information is available on the suspected properties of *L. garcinnii*.

7.4   Individual variation and quality

A priori, one would not expect all Oman's immigrant oryx, or even the Oman-born animals, to be equally successful or adapt universally to the rigours of the desert. The length and number of inter-calf intervals of different

Table 7.8. *Details of stillbirths and one premature birth in 1983–4*

Oryx	Date mated (day/month/year)	Expected date of calving (day/month/year)	Days early	Comments
Hadya	4/04/83	24/12/83	24	Calf small and weak at birth, but survived due to intense parental effort. In herd 1 in desert
Talama	9/04/83	29/12/83	22	Stillborn, fully formed. In herd 1 in desert
Alaga	20/05/83	8/02/84	41	Stillborn, eyes unopened. In herd 1 in desert
Farida	8/06/83	27/02/84	28	Calf never found, presumed stillborn. In herd 2 in enclosure

females suggested this, but two approaches are now presented to describe individual response to desert conditions. The first is quantitative through assessment of the reproductive rate and success of females. The second is for the most part descriptive, using the whole range of management observations on animal appearance, condition and performance as indicators of adaptation and welfare. The implications of these observations for the number and selection of further animals for re-introduction are discussed later.

Fig. 7.5. Matrilines in Oman's oryx.

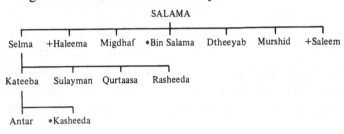

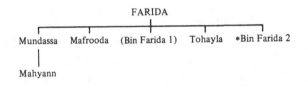

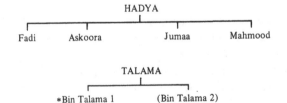

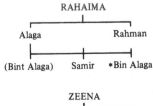

( ): stillborn
*: died before 30 days
+: died after 30 days

7.4.1   *Matrilines and success*

Six immigrant females produced at least one live- or still-born calf each in Oman (Fig. 7.5). The six females were responsible for 32 births. The contributions of the lines were very unequal. Salama's line was responsible for 13 out of 32 births because she had seven calves, and both her daughter, Selma, and grand-daughter Kateeba, bred before mid 1986. In contrast, two females, Talama and Rahaima, each had only two calves in their lifetimes. Zeena is younger, alive and breeding.

When mortality is combined with birth statistics (Table 7.9), the differences between lines are exaggerated. The largest matriline had 85% survival of calves to 30 days old, while Talama's only two calves were stillborn or died at 3 days. Survival to this age was 100% in the lines of Hadya and Zeena, with four and two calves, respectively.

The main factor behind this variation in calf production between lines was not an adjustment in the inter-calf interval. Calculation of the intervals between the calves of each female conceived in Oman removes any effect of the variable period between arrival in Oman and first conception (Table 7.9 and 7.2.6). The three females of Salama's line showed consistently short average intervals between 9.0 and 9.8 months. Two successful females, Hadya

Table 7.9. *Calf survival and intervals between calves conceived in Oman for females and matrilines*

Matriline	Female	No. born	Surviving 30 days		Av. interval between calves (months)
			No.	%	
Salama		13	11	85	—
	Salama	7	6	86	9.8
	Selma	4	4	100	9.0
	Kateeba	2	1	50	9.2
Farida		6	4	67	—
	Farida	5	3	60	11.2
	Mundassa	1	1	100	(> 17 mo. on 31 July 1986)
Hadya	Hadya	4	4	100	11.3
Talama	Talama	2	0	0	10.4
Rahaima		5	3	60	—
	Rahaima	2	2	100	(Died 10.7 mo. after birth of only calf conceived in Oman)
	Alaga	3	1	33	13.5
Zeena	Zeena	2	2	100	13.1

mo., months.

and Zeena, had intervals of 11.3 and 13.1 months, while two unsuccessful animals, Talama and Alaga, had intervals of 10.4 and 13.5 months. The failure of Mundassa to calve for the second time by 17 months after her previous calf is anomalous. Talama's line ended with her death in the desert, while Rahaima's continued after her death through her daughter Alaga, only one of whose three calves survived beyond 30 days.

7.4.2   *Correlates of individual success*

In long-term studies of ungulates, weights at birth can be related to the climate during the calving season. Changes between summer and winter coat vary with the reproductive status of females and the density of animals (Clutton-Brock, Guinness & Albon, 1982). The oryx in Oman presented fewer opportunities, for calves were not weighed due to the risk of abandonment by the mother, but observations on calf size and behaviour, especially during the first two months of life, indicated the hardiness of each calf. Similarly, adult coat colour, texture and their changes through the year (2.4.1), the behaviour and visible condition of each oryx all combined to allow a rough assessment of adaptation to prevailing conditions. These indicators are corroborated here with poor performances for certain 'low-quality' animals.

Talama

This female was in the first group of immigrants from the USA. Her breeding record was poor, with one calf surviving to 2 days and the next stillborn, and when she died (Table 7.7) at 69.4 months, her line became extinct.

Both in the enclosure and desert, her summer coat retained a tinge of yellow, never achieving a totally white colour, and her body condition was never as good as other females. When herd 1 made the long journey from range 2 to 3 (Fig. 6.5a, p. 131), 16 days before she calved for the first time, Talama lagged far behind the herd, walking listlessly while relatively emaciated. Although she conceived shortly after the birth and death of her first calf and again after her second, she again lost much condition during her third pregnancy, and in her last month of life often trailed. At death her femur marrow was 93% moisture, indicating very poor body condition (Sinclair & Duncan, 1972).

Her cause of death was confirmed as botulism, and she had a long history of scavenging textile material and gazelle carcasses. This habit could either have been picked up in captivity in the USA or was the 'pika' of animals that are mineral-deficient (Hall, 1977). Her faecal nitrogen concentrations were on average 32% higher than the rest of the herd, whether in the enclosure or desert, on supplements or grazing or a mixture (Table 7.10). If this difference roughly measured the inefficiency of her nitrogen metabolism, it would explain her poor condition on diets which were at or above maintenance standard for the other oryx. Thus, she was not fit enough for the desert environment.

Rahaima

This oryx was also an immigrant but she and Museba were bred in Gladys Porter Zoo, Brownesville, Texas, from Phoenix Zoo stock. Both were small animals of conformation distinctive and different from that of the San Diego animals (5.2.3).

Although able to maintain condition better than Talama, Rahaima was an inveterate scavenger and chewer of bones. Her faecal nitrogen levels were always in the range of values of her herd. Her breeding record was also poor, with 29 months between the birth of her first calf, conceived in the USA, and her second. She was pregnant with her third calf at her death. The rearing of her second calf, Rahman, and of a similarly aged calf by another female illustrate the selective pressures imposed by the desert. Rahman was born in range 7 (Fig. 6.4c) in November 1983. Herd 1 had found green but drying grazing in this area, but had not drunk free water for three months. Rahaima lactated through the cool winter months despite not drinking. Through early 1984 air temperatures rose and grazing quality declined, leading up to the herd's return to the release area. Rahman was not seen sucking after late April

Table 7.10. *Faecal nitrogen levels of herd 1 and Talama, with some duplicate measurements*

Date	Faecal nitrogen, %				Comments
	Talama	Herd 1	s.D.	n	
Nov. 1981	1.48	1.24	± 0.10	9	In enclosure, pre-release[a]
Apr. 1982	1.64	1.20	± 0.16	5	In wild, no supplement
Jul. 1982	2.80	2.00	± 0.38	6	In wild, no supplement
Jun. 1983	1.32	1.09	± 0.10	8	In wild, no supplement
	0.96				
Apr. 1984	1.22	0.86	± 0.07	10	In wild, no supplement
Aug. 1984	1.20	0.81	± 0.06	7	Poor grazing with light
	1.26				supplement
Nov. 1984	1.38	0.98	± 0.18	7	Poor grazing, no supplement
	1.27				

[a] $n = 9$, excludes Sajba in very poor condition; faecal N 1.63%, 1 month before death.

when he was 6 months old. His continuing attempts to suck showed that weaning was caused by Rahaima's inability to produce milk, for her udder had regressed. Rahaima first returned to Yalooni for water in early June, and within 10 days of her drinking regularly, Rahman was sucking again as his mother's udder re-developed. He was not finally weaned for another 2 months, by which time he was 9 months old. Comparison with Hadya's calf, Askoora, born into the same herd less than a month later showed that Rahman's growth may have been permanently stunted by his early weaning on to an inadequate diet. It is unlikely that Rahman will ever be large enough to compete effectively to become a breeding male. Thus, Rahaima's line will persist only through Alaga, her first calf, who has a poor record for calf survival (Table 7.9).

Nafis

He was a San Diego animal, released into the desert in herd 1, where he was the subordinate of four males even though the second oldest. After the two herds met (6.8), Nafis changed to herd 2, where two males were dominant to him. While in herd 1 he often took the role of leading male when the herd was moving large distances but efforts to assert himself were few in the 48 months in the desert before his death: he managed to consort briefly with Rahaima and then Talama in range 7. Later, he separated Talama from a subgroup of herd 1 and held her for 5 days during an oestrus, during which he mated her repeatedly but unsuccessfully. He did not squat-defaecate during this consortship. Finally, for three months up to one month before his death, he lived with a subgroup of four oryx from herd 2 (6.7). During this period, he never squat-defaecated nor herded the females, suggesting little inclination to assert any dominance. An unmended broken rib found on his autopsy probably resulted from a low-intensity tussle with another male two to three months before his death. This wound may have caused him to join the female group with its lesser chance of interactions with other males.

Through his time in Oman, Nafis' coast was always longer and coarser than in any other animal, and as his condition deteriorated through his last 6 months, the difference became more apparent. He displayed another symptom of inability to adapt to local conditions. Starting in April 1985, 9 months before his death, he frequently developed a temporary submaxillary oedema. Although a few other animals of both herds had the same condition occasionally through the early summer, Nafis showed it frequently. The oedema, which lasted up to 72 h, was invariably associated with drinking 12–24 h previously. By September 1985 he sometimes did not walk into water with his herd as if deliberately to prevent an oedema from developing.

Paradoxically, the oedema would have been avoided more effectively by taking small drinks more often as it was probably due to poor peripheral circulation. His condition allowed tissue fluid to accumulate at a time when the body fluids had been increased through drinking. There were many possible reasons for Nafis' poor circulation, but its consequences were further indication of his inability to thrive under desert conditions. He died without leaving any offspring.

Hadya

This female oryx was a large animal and a successful breeder (Tables 7.3 and 7.9), producing four large calves, all of which survived, at short intervals. Subjectively, her mothering behaviour was the best of all the breeding females. However, she had a nervous temperament, which required her release from the pens into the enclosure only 2 days after arrival from the USA (5.2.5). In herd 1 she was calm and approachable, but tended to panic when separated from the herd in the desert. Efforts by patrols under these circumstances to keep her under observation or direct her towards the herd had to be done more circumspectly than for any other oryx. She died when alone with her 5.5 month calf in range 12, after being alarmed by an approaching vehicle whose driver was unaware of her position. The autopsy showed she had recently aborted a small foetus but she was in excellent condition. The cause of death was surmised as stress-induced *Salmonella* or botulism. Thus, a behavioural short-coming, the tendency to panic when away from her familiar herd, was responsible for the death of an otherwise good, desert-adapted female.

Apart from Hadya, these other three immigrants were clearly 'low-quality' oryx. As their genetic representation in the oryx born in Oman has been very low, their genes have been largely lost. As described above, Rahaima's surviving descendants, Alaga and Rahman, will contribute little to future generations. Thus, the oryx population's fitness is presumably increasing.

Mustafan

This male posed a dilemma because of his social role. He became herd 2's dominant male in September 1985 (6.8). His herd contained three proven breeders and three females of breeding age. After 17 months with the herd, no female had produced a calf by Mustafan and none appeared pregnant. Intervention was decided upon because of the demographic implications of his presumed infertility. He was immobilised in February 1987. As electro-stimulation yielded no semen ejaculate, a closed castration was carried out although his testes had largely regressed. He was negative for *Brucella* sp.

From his subsequent behaviour he retained his dominance, but his inability to fertilise females was not serious as the original herd 2 split into smaller units which were widely spread throughout the desert (6.7; T. H. Tear, personal communication 1987).

In contrast to the immigrants, oryx calves born in Oman were subject to more natural selective pressures from birth. Differences in their early behaviour were possible indicators of future performance. Saleem, who was killed by ravens at 1.7 years was considered from soon after birth to be weak, which showed in more separations from the herd and less ranging by him and his mother, Salama. Salama's previous calf, Murshid, was born in early May 1985 in very hot weather. By 6 weeks he was still left lying out rather than following his mother to the feed troughs each afternoon like other calves of his age. Often when on the move, he lagged and hyperventilated. At 3 months old, Murshid was noticeably smaller than his age peers in herd 2. The gradual change of his calf coat to white was also delayed. These observations suggested that Murshid would not be a large adult, and consequently was less likely to be a successful male.

7.4.3   *Phenotypic variation between oryx in Oman*

Strong family resemblances ran through Oman's oryx, as has been observed in other closely studied ungulates (Clutton-Brock *et al.*, 1982). They were evident in body size, body conformation such as the profile of the back, the degree of inturning of the hind legs, and especially in horn length and shape. Different lines also had characteristic development of the black flank stripe and intensity of the dark markings. The two animals from Brownesville Zoo differed remarkably from the San Diego stock (5.2.3).

Amongst the calves born in Oman, the two hand-reared ones were both small adults, but, at least in the case of Selma, this was no handicap to her breeding performance (Table 7.3). Such small females produced average-size calves, which quickly overtook their mothers in size. There was considerable variation in the rate of calf growth in Oman, and this was most obviously due to grazing conditions during early life. An uncertain food supply in the desert during growth is likely to cause a wide range of animal sizes for age. This may influence the dynamics of the growing population if fertility is related to body size, as in red deer (Albon *et al.*, 1986).

In view of the effect of nutrition on growth rate, oryx born in Oman were likely to be smaller on average than those in zoos, where the plane of nutrition can be consistently higher and certainly less subject to seasonal variation. Thus, under captive or managed conditions, the animals more closely attain their genetic potential size, as in red deer (Suttie *et al.*, 1983). Oryx born and reared in Oman also differed phenotypically in one respect. They had a more pronounced hyoid structure than the immigrants. As the hyoid is involved in tongue action during grazing, it may develop differently in oryx which have grazed since weaning from those eating only cut forage in captivity. If so, immigrant oryx may never match the adaptation of desert-born animals to their natural environment.

7.5   Genetic diversity and inbreeding

The proximate causes of death discussed above showed little pattern, while the observations on individual performance in the desert were only qualitative. However, the existence of accurate pedigree data on each oryx in Oman, going back to the World Herd founders, allowed examination of genetic diversity in Oman's population relative to the parent, USA population. It also enabled calculation of the extent to which each immigrant oryx was inbred, and this could be maintained subsequently for Oman-born oryx because of their known parentage.

Inbreeding in small captive populations of ungulates has severe effects on juvenile survival (Ralls, Brugger & Ballou, 1979). In captivity, scimitar-horned oryx calves suffered a higher mortality if their mothers had been mated by a related male than by an unrelated one. Any adverse inbreeding effects might be of critical importance to the Oman project, requiring action to manipulate the breeding system or as a basis for objective identification of future immigrants.

7.5.1   *World Herd founder representation in Oman*

The oryx imported into Oman from the USA were selected for a certain age structure and sex ratio, but substitutions were unavoidable after veterinary testing (5.2.2). From the genetic point of view the immigrants were a random selection. Comparing the proportional representation of World Herd founders in the USA oryx of the 1984 studbook with that of the oryx in Oman in 1985 (Mace, 1986), nine founders had between 5% and 19% representation in each population (Fig. 7.6). This indicated that the re-introduced population had a broad genetic base, which had not narrowed over the years in Oman. However, the contributions of individual founders differed in the two populations, and American animals descended from those which were underrepresented in Oman would be obvious choices as further immigrants (9.5.1).

7.5.2   *Inbreeding coefficients*

The pedigree of each oryx imported into Oman from the USA could be traced to the World Herd founders with almost total confidence. Assuming that the founders were themselves unrelated, the degree of relatedness could be calculated for any pair of oryx. The inbreeding coefficient for any individual is half the degree of relatedness of its parents, and it represents the probability that two alleles at a locus in the individual are identical by descent from a common ancestor. The values for each oryx in Oman are shown in Appendix 3, and were calculated from studbook data by Dr Georgina Mace.

The distribution of grouped inbreeding coefficients shows that 12 out of 18 immigrants were not inbred at all (Table 7.11). The most inbred oryx were Jadib and Museba, each with a coefficient of 0.125. Conventionally, at coefficients above this value, equal to those of offspring from a first cousin

Fig. 7.6. World Herd founder representation in (*a*) Oman's oryx in 1985, (*b*) North American oryx in 1984. (From Mace, 1986).

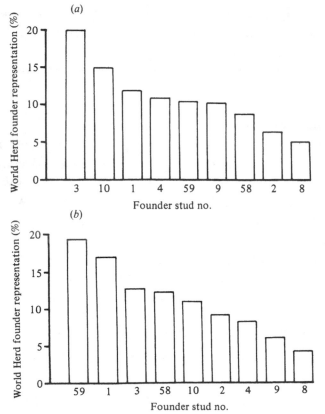

mating, inbreeding effects might be apparent. Thus, the overall level of inbreeding in the immigrants was low.

The distribution of inbreeding coefficients of the 32 oryx born alive or dead in Oman was different (Table 7.11). Whereas the average coefficient of the immigrants was 0.028 (s.D. ±0.048), that of the Oman-born oryx had more than doubled to 0.065 (s.D. ±0.044). Of the increase in the average, 16% was contributed by one animal, Alaga, whose inbreeding coefficient was 0.242. Her parents originated from Brownesville Zoo and were themselves half-siblings, while a few oryx from Phoenix occurred repeatedly in the three generations between the World Herd founders and Alaga (Fig. 7.7).

7.5.3   *Inbreeding, survival and performance*

The eight immigrants which had died by 31 July 1986 were a small sample but, on a percentage basis, their inbreeding coefficients reflected their proportions in the population (Table 7.11). However, it is not justifiable to

Table 7.11. *Inbreeding coefficients and mortality in immigrant and Oman-born oryx*

	Inbreeding coefficients					
	0– 0.025	0.026– 0.050	0.051– 0.075	0.076– 0.100	0.101– 0.125	0.226– 0.250
Immigrants						
No.	12	1	2	1	2	0
%	67	6	11	6	11	
Dying by 31 July 1986						
No.	5	0	1	1	1	
%	62.5	0	12.5	12.5	12.5	
Names of those dying:	Kadil Talama Sajba Malak Hadya		Nafis	Rahaima	Museba	
Born in Oman						
No.	4	10	8	5	4	1
%	12.5	31.25	25	15.6	12.5	3.1
Stillborn or dying < 30 days						
No.	1	0	2	3	2	0
%	25	0	25	60	50	0

conclude that survival was independent of inbreeding. Of the six immigrants with inbreeding coefficients greater than zero, three died. Two, Nafis and Rahaima, were obviously low-quality animals (7.4.2), and Museba's performance was indifferent in that he was an unaggressive oryx and, consequently, left no offspring except Alaga, who had been conceived in captivity in the USA.

Apart from these three oryx, five other immigrants with coefficients of zero died. The deaths of Kadil, Sajba and Malak were due to old age or accident (Table 7.7). Of the others, Talama was clearly ill-adapted to the desert, while Hadya was a high-quality oryx apart from the fatal flaw in her temperament (7.4.2). Thus, it seems that poor survival or performance in immigrants may

Fig. 7.7. Pedigree of Alaga, with studbook number of each oryx.

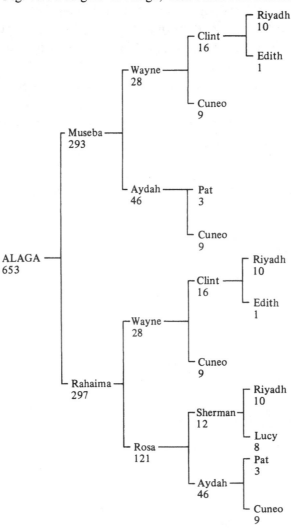

have been related to the level of inbreeding but also occurred in non-inbred oryx which could not adapt to the desert.

The possible effects of inbreeding on survival in Oman-born oryx were examined by survival to 30 days from all births, whether live-born or stillborn. This measure discriminates best between deaths related and not related to inbreeding (Templeton & Read, 1984). Out of a total of 32 live and stillbirths in Oman, three calves were stillborn and a further five died within their first 30 days. Comparing the coefficients of the 32 born, the average inbreeding coefficient of the eight dying was 0.084 (s.d. ±0.016), and that of the 24 survivors was 0.061 (s.d. ±0.047) (Table 7.11). Such a difference with small sample sizes was not significant because of the effect of Alaga surviving with a coefficient of 0.242 and Alaga's third calf dying despite a coefficient of 0.008. However, from survival in each inbreeding coefficient class, the chances of survival appear to drop rapidly as the inbreeding coefficient exceeds 0.076 (Fig. 7.8). This conclusion has to be qualified by the survival of the sole animal in the 0.226–0.250 class, but her early history suggested that she would not have survived without human assistance (5.2.8). Thus, these data suggest that mortality in Oman-born oryx was related to their inbreeding coefficients. The effect was apparent at coefficients far below the 0.125 level at which such effects might start appearing in captive populations. This might reflect the more severe selective pressures to which the desert-born oryx calves were

Fig. 7.8. Numbers of oryx born in Oman, including still-born, dying before, or surviving beyond, 30 days old, with inbreeding coefficients.

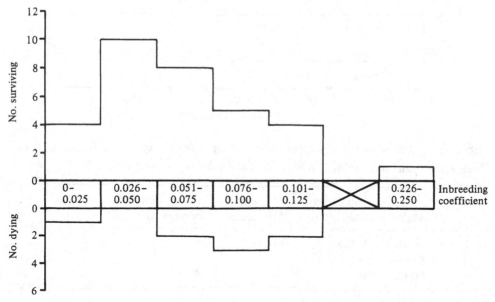

subject. Its implications for the descendants of the oryx currently in Oman and for the selection of future immigrants are discussed below.

7.5.4   *Inbreeding, matrilines and patrilines*

Calf survival varied between matrilines (7.4.1). The percentage of calves dying in each matriline before 30 days is plotted in Fig. 7.9 against the average inbreeding coefficient of all the calves of the line and their offspring. Despite the uneven variances around the averages, there is a general trend of lower survival with increasing inbreeding level for the six lines.

The inbreeding coefficient of any calf depends on the degree of relatedness of its parents. If both parents are inbred but not related, their offspring will not be inbred. These two measures need not be correlated, as the three males, who between them produced 30/32 calves born in Oman, show (Table 7.12). Average relatedness for each male is calculated as the average of his relatedness with each female by whom he had at least one calf, but it is not

Fig. 7.9. Average inbreeding coefficients, with 95% confidence limits, and mortality within 30 days of birth for calves of six matrilines. The number of calves in each matriline is given in parenthesis.

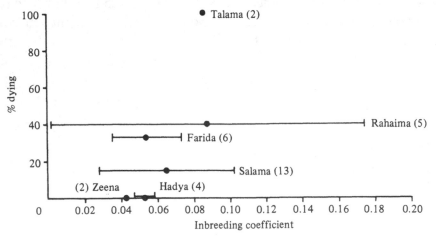

Table 7.12. *Inbreeding coefficients and average degree of relatedness between three males and the females with which each bred*

Male	Inbreeding coefficient	Relatedness	No. of females bred with
Jadib	0.125	0.163	6
Lubtar	0	0.118	4
Hamid	0	0.063	4

weighted by the number of calves by each of these females. Jadib had an inbreeding coefficient of 0.125 and an average relatedness of 0.163 to the six females who had his calves. The other two males were not inbred at all, but, while Lubtar had a relatedness of 0.118 to his females, Hamid's was only 0.063. The effect of this on survival of their calves shows as a linear relationship (Fig. 7.10) between calf survival to 30 days and the average inbreeding level for the patriline, subject to the caveat that this is based on the results from three lines only. This relationship suggests that, to avoid calf mortality due to inbreeding-related effects, breeding males should be selected for their offspring to have inbreeding coefficients as low as 0.010. Thus, in Oman, future immigrant males should be chosen, at least, for their unrelatedness to the females of their expected herd.

7.6   Discussion

A larger sample of births in Oman is needed to confirm whether the Arabian oryx has the same breeding season at both 33° N in the USA and 20° N in Arabia. If so, it suggests a genetic base to breeding season. Introduced populations of reindeer retain the breeding seasons of their source areas, disregarding any 6 month reversal due to change of hemisphere (Leader-Williams, 1988). This suggests a genetic basis to breeding timing which will only adapt slowly in translocated animals. The coincidence of breeding seasons in both Oman and the USA at 33° N indicates, therefore,

Fig. 7.10. Average inbreeding coefficients, with 95% confidence limits, and mortality within 30 days of birth for calves of three patrilines. Number of calves by each male in parenthesis.

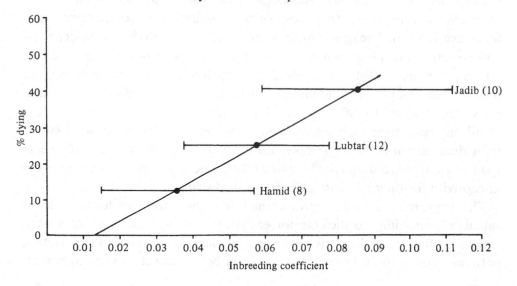

that the zoo animals retained the season of the aboriginal population. Although long daylength seems to be a major determinant of seasonality, enough births occur in non-peak months to infer that other factors operate. These are likely to include grazing quality or body condition to give wild populations a prompt response to improved conditions after rain. The generally greater age of first breeding in the desert than in captivity (7.2.5) implicates body size or condition due to the superior diet of zoos. Only an occasional, wild female grows up under lush, desert conditions and breeds early. Although fecundity in captivity may be greater, the Oman population increased as fast as the World Herd (7.2.1). Against this pattern, the delayed breeding in females arriving in Oman around the expected age of sexual maturity is surprising, because daylength change was not great, and the species appears to be totally unstressed by crating and air travel (5.2.2).

Both in captivity (3.5.2) and in the wild, oryx suffer a low mortality after the age of 6 months, suggesting either a robust species or a non-selective environment. However, the differential mortality between lineages shows that natural selection in the desert's extreme conditions operated swiftly on animals that were for the most part only third or fourth generation captive-bred. The lesson for other re-introductions is that deaths must be expected amongst captive-bred animals fairly soon after release. But they need not be a sign of failure, especially if related to pedigree line. Interestingly, it looks as if the surviving offspring of low-quality oryx are themselves likely to leave few offspring. Thus, the genetic base in Oman is narrowing, indicating the need for an increase in genetic variability.

When the oryx were well known as individuals, correlates of individual performance were obvious. Although qualitative and largely based on anecdotal observations, they complement analysis of comparative performance between lineages. Observations such as the deduced influence of diet structure on the anatomy of the upper oesophagus could be used to introduce more realistic conditions in captive-breeding programmes to produce the best possible individuals for re-introductions. While in this case early diet appeared to influence hyoid development, captive conditions must avoid imposing their own set of selective forces (Frankham *et al.*, 1986) if individuals are to be fit for release into the wild after the expected 200 year-span of captive-breeding (Soulé *et al.*, 1986). The oryx, in comparison, were in captivity for only 20 years before re-introduction.

The oryx re-introduction project had the good fortune to be able to calculate inbreeding coefficients for each oryx because most animals were descended from a small captive nucleus a few generations previously. Hence, patterns, which would never otherwise have been suspected, were apparent.

The relationship between survival and inbreeding level was clear at unusually low levels of inbreeding (Ralls *et al.*, 1979), which might have been due to relaxed or very different selective pressures in captivity. The infrequency of close inbreeding in wild populations (Ralls, Harvey & Lyles, 1986) suggests that inbreeding coefficients are correspondingly low. Although deliberate inbreeding in captivity can result in viable animals if the deleterious recessive genes are purged (Templeton & Read, 1984), and some wild populations apparently thrive with no genetic variation (O'Brien *et al.*, 1983), rising inbreeding levels could endanger the oryx re-introduction. Sound, early captive-breeding policies nursed the small Arabian oryx population through a genetic bottleneck (3.6), and the Oman herd fortunately had a broad genetic base (Fig. 7.6). However, its continued success may depend heavily on increasing genetic variability and reducing near-kin matings to minimise inbreeding levels.

The success of oryx herds in the desert does, indeed, depend on their members, but it is clear that this success is only achieved at considerable cost to some animals and lineages. The implications for the re-establishment of a viable oryx population are discussed in Chapter 9, after analysis of the human side of the project, and its bearing and contribution to the re-introduction.

8

The Harasis and the re-introduction project

8.1 Introduction

Many aspects of the behaviour and performance of the re-introduced oryx were shaped by their interactions with the local people of the re-introduction area. Indeed, most of the observations in previous chapters were made by the project's Harasis rangers. This chapter examines the impact of the project on the local people and area, to assess its social and economic influence, and its relationship to the spatial patterning and distribution of the tribesmen. The areas of conflict and co-operation between the project and people are described with a basic analysis of prevailing attitudes towards the oryx and their return. Finally, as the way of life of the Harasis has changed dramatically over the last 15–20 years, an attempt is made to predict the nature of future changes and their impact on the environment of the people and oryx.

8.2 Employment on the project

8.2.1 *Staff and organisation*

Most project staff were Omanis (Table 8.1). They included Harasis *bedu* and villagers with homes 300 km from the re-introduction area. A smaller proportion was expatriate. Between 1980 and the end of 1986, the establishment grew (Table 8.1). Total staff increased from 28 to 41, and the proportion of Harasis increased from 54% to 63%.

Single accommodation for all staff was provided at Yalooni, and an Omani mess run at cost for staff (see below). Monitoring the oryx required staff on duty every day of the year. Because of this, and the distance to the villagers' homes, staffing operated on a flexible shift system with a block period of 7 days each month off duty for each Omani. Rangers were technically on stand-by duties in camp when not on patrol. In addition to these 7 days each month,

202

each man was entitled to all Public Holidays, usually added to a week at home or as staff schedules permitted, and 30 days' paid annual leave.

Project transport was provided for everyone once a month for the return journey between camp and home, but many Harasis preferred to use their own car and a petrol allowance was given in lieu. Uniforms were provided each winter and summer, consisting of sandals, the traditional robe with a coloured head-dress, and two blankets (Plate 4b). Each ranger had a 0.303 rifle, bandolier and ammunition for which he was responsible. Staff families living near Yalooni were allowed to draw 400 l of water a day for their homes. This concession enabled employees to go on patrol or on vehicle missions of several days without constantly worrying that their homes were waterless.

8.2.2   *Salaries and costs*

A ranger's salary was made up of a basic sum and an allowance for his rations, a fixed sum for overtime and another for desert-living. In 1986, a ranger who had worked for 5 years, receiving annual increments of 5% of his basic salary, earned RO.225 each month (RO.1.000 = US$2.55 in 1986). RO.25 was deducted at Yalooni for the cost of rations in each month the person was in camp, and the small surplus which accrued was returned as a rebate each December.

This salary must be seen in the context of local conditions. The Harasis now practise a form of motorised pastoralism. Even if a family maintained only

Table 8.1. *Oryx project staff establishment in 1980 and 1986*

	1980			1986		
	National		Expatriate	National		Expatriate
	Harasis	Other		Harasis	Other	
Rangers	10			17		
Drivers	4			6		
Workshop	1		1	2		2
Kitchen and camp		8	2	1	6	3
Management and administration			2			4[a]
Total		28			41	

[a] Includes one clerk in Muscat.

one pickup in roadworthy condition, its licencing and insurance cost the equivalent of RO.12 per month (Table 8.2), while two drums of fuel a month probably underestimates actual consumption. An average family's food costs a minimum RO.60. The biggest item of expenditure is concentrated animal feed. Although the household milk camels might be supplemented almost constantly, feeding of all camels, goats and sheep started about 6 months after rain. An average household would then use one 50 kg bag of concentrates daily. As concentrate use depended on the rain pattern on the Jiddat-al-Harasis (4.4), periods without high feeding costs were short. Only then was there surplus cash for items such as clothing, vehicle spares, the running costs of second cars or water-carriers, the replacement costs of vehicles, or frame and tarpaulin shelters.

These basic figures show the significance of the family's livestock for milk, meat and cash from sales to its economy. However, a detailed and systematic study of contemporary Harasis household economics showed that some families sold no livestock (Chatty, 1984) and few probably earned more than a few hundred rials a year through livestock sales. The Harasis' best hope for a significant capital gain was through selling a young camel with racing potential for RO.500–3000. For most families in most years, a salary of RO.200 per month had to be augmented by the products of pastoralism and, preferably, by a second salary in each family. Combined family incomes are now often insufficient, resulting in indebtedness to wealthier members of the tribe, village shop-keepers and car agencies. One view is that debt management within the tribe has replaced livestock management (Chatty, 1984).

8.2.3   *Strategies for employment and pastoralism*

To combine work on the re-introduction project while continuing as largely absentee pastoralists, the Harasis had several strategies designed to maximise benefits or minimise costs, depending on circumstances. Four main strategies were apparent.

Table 8.2. *Minimum monthly expenditure in average-sized Harasis family under dry conditions*

Item	RO.
Animal feed: 30 bags at RO.3	150
Family rations	60
2 drums petrol (400 l)	52
1 vehicle tax and insurance	12
Total	274

(1) After rainfall that would produce fresh grazing, almost all families moved with their livestock to the greening area. The reduction of feed costs outweighed any disadvantage of greater distance between home and place of work. Families that did not follow the majority were either those who had settled to cultivate or who had to remain within reach of hospital because of chronic sickness.

   The remaining three strategies related to periods when grazing was patchy or was of relatively uniform but declining quality over large areas, or was non-existent.

(2) Families moved to within commuting distance of Yalooni. Wives or children drove in from up to 15 km each day to take water. Husbands went home to help with feeding or milking camels each evening when on stand-by duties rather than on patrol or away from camp. Such families tended to move less often, and the consequent loss of animal production and higher feed costs were offset by the guaranteed water supply, the more frequent presence of the husband, and fewer problems associated with changing living site.

(3) Families living further from Yalooni depended on water supplied by a government tanker, operated by a Harsusi. The latter lived within reasonable reach of a permanent water-well, and his client families had water delivered less often with increasing distance from his home. Under these conditions an employee of the oryx project might visit home every 3–4 days if within 50–80 km, taking full drums of water with him. His travelling costs were higher on each visit, but his family's greater mobility might mean lower feed costs while other family labour, such as daughters, allowed his less frequent presence. The family's costs increased dramatically, absorbing up to half a salary, if it had to transport water while the government tanker was out of action (Chatty, 1984).

(4) When families lived further than 100 km from Yalooni, the employee visited home rarely more than once a month for his period off duty. Such families comprised those which were either highly mobile in the search for grazing, although this category was small, or those who traditionally ranged through more distant areas. Families increasingly settled near or moved around a favoured water-well, especially in the south-west Jiddah (Fig. 4.3), while one ranger started to cultivate a small area 250 km from Yalooni, left in the charge of an adult son. Employees in this category avoided travel costs as far as possible.

To large degree, these strategies were stages on a continuum, and one family might change between them depending on conditions. Few did not respond to rain as in (1) above. Under other conditions each family tended to stay with one strategy because of the size, mix and importance of its livestock holding, the total of incoming cash, the family's vehicle ownership, the extent of family labour and closeness of relationship to a shaikh or tanker operator. Temporary or fairly permanent larger economic units than a single nuclear family were created if a man had adult sons who either worked in the family camp or for cash elsewhere, or if two brothers, each with a family, combined so that one worked for a salary while the total livestock and water supply were the concerns of the other.

8.2.4   *Administration and management in practice*

While the rules concerning periods on and off duty for Harasis staff were simple and judged fair over the years, their implementation and enforcement were inevitably tempered by local circumstances. Few *bedu* completed a month without having adequate reason for being away from the project for more than the stipulated number of days. Modern amenities such as medical attention were desired, but all services were still sparse in the large area of central Oman. A project ranger might be obliged to drive a family member 150 km to Haima clinic, and then spend 5 days in attendance there while the relative had one daily injection. The lack of in-patient facilities increased the need for family helpers for the sick. As goat herds became less efficiently watched and managed (4.8.2), straying increased and staff appealed urgently to be allowed to leave to track up missing animals before they wandered too far, were predated or died of thirst. It was hard not to allow someone to deliver water to his home when the usual tanker was known to have broken down. If a peak of camel and goat births occurred under dry conditions, only skilled and intensive management could ensure suckling and bonding. With camels, only a man can cross-foster a young calf on to a second female, whose newborn male calf would be slaughtered. This was an important means of rearing one good female camel and increasing the household milk supply.

These examples demonstrate the type of conflict of interest and the resulting, multiple demands on a man's time when the oryx project operated within the tribal area. News of problems at home invariably spread fast to Yalooni. Such conflicts were inevitable when the project tried to ensure that paid employment was compatible with continuing pastoralism, and, in recognition of the situation, the staff establishment contained a little spare manpower. However, genuine reasons for extra absence from camp had to be carefully checked and distinguished, leading to very time-consuming staff management at all hours of the day.

8.3   Informal benefits from the project

Completion of the oryx project camp in 1979 preceded that of Haima (4.8.1, Fig. 4.6), making it the only settlement of any sort west of Duqm for 500 km. The Harasis had been used to travelling 300 km north-east to the nearest village to buy petrol or even to have a split water tank welded. Yalooni at once became an unofficial community centre and destination for any Harsusi to seek assistance over minor problems. This might be given if it in no way interrupted the running of the oryx project, and if the real cost of the help to the project was almost insignificant.

The scope and types of assistance most commonly rendered are shown in Table 8.3. Most forms cost the project little but would be expensive to the recipients if they had to find them elsewhere. Outsiders were entertained to

Table 8.3. *Main informal benefits from Yalooni camp*

Category	Benefits
1. Workshop	Ramp for oil changes, repairs
	Puncture repair equipment and compressed air
	Use of tools on site
	Technical advice on repairs, fault diagnosis
	Welding of water tanks, tent frames
	Loan of vehicle spares against replacement
2. Camp	Mosque, especially for Friday prayers
	Bathroom
	Coffee, meals for passers-by
	Petrol: 20–40 l as emergency assistance, up to 200 l on loan
	Provision of air-conditioned cabin for women by day for holy month of fasting, Ramadan
	Basic medical assistance
3. Labour	Many, such as helping to track camels or load them into pickup trucks; providing labour to dig a grave within prescribed time
4. Communications	Radio link to Royal Oman Police or to conservation office in Muscat concerning people in hospital or for urgent messages
	Informing oil company when diesel delivery needed at Al Ajaiz well
5. Information centre	Information on location of people, homes, where camels seen etc

meals at the expense of the *bedu* mess, and the goodwill engendered resulted in project drivers, other than those on oryx patrol, never being rationed when on duty on the Jiddah, for they enjoyed reciprocal hospitality.

All *bedu* peoples are resourceful and alert to the potential of a new situation: staff soon realised the advantages of Yalooni for keeping some camels. This meant that milk was not only a luxury during periods at home. In addition, the permanent presence of some men in camp guaranteed the feeding and watering of any camel returning while its owner was away on duty or on leave, and such favours could always be reciprocated the same month. Renewed interest in camel racing stimulated further investment and effort in camel management at Yalooni, without detriment to the project.

The conventions surrounding assistance were not abused, and no one borrowing 200 l of petrol ever failed to return it at his convenience. Through this role for the camp, the project accumulated a fund of non-financial obligations that could be called in for return at any time. This rarely happened, but the response could be vigorous: when, in 1982, a female oryx and her 17-day calf, the first born in the wild, disappeared from Yalooni in a sand-storm and were found safe only after 3 days, men from about 30 families searched for their tracks through likely places over a vast area. When the oryx were found, not one claimed any petrol or compensation for his efforts. On the same occasion, an old woman, walking with her goats, followed for several kilometres a set of donkey tracks, a rarity on the Jiddah, thinking they were those of the oryx.

8.4   Land-sharing arrangements between the project and Harasis

The entire central deserts of Oman are common land, although areas are nominally recognised as those of specific tribes for purposes of grazing. Even after the Jiddat-al-Harasis is declared a national nature reserve (IUCN, 1986), *bedu* will continue to live in it, and the re-establishing oryx will have to share their range. However, for the sake of promoting the success of the re-introduction, two restrictions on settling and development were made in 1979.

During camp construction the Harasis agreed that no families would settle in the 10 km² of *Acacia–Prosopis* woodland and perennial grassland around the enclosure (Fig. 5.3). As the Yalooni *haylah* is the largest on the central Jiddah, its browse, shade and shelter made it a most important area of permanent resources for the Harasis. This agreement not to settle, made before the project could offer anything in return, was a considerable sacrifice and token of good will, and the assistance described above was rendered in this context. In addition to keeping the release area undisturbed for the oryx, the absence of *bedu* camps lessened the possibility of disease transfer or exchange between oryx and livestock. It was also intended to ensure some grazing in the immediate release area. This was achieved by protecting the grazing from goats between 1979 and the first release in early 1982. In subsequent years, *bedu* camps were rarely less than 3 km from Yalooni camp, and their distribution pattern was not uniformly encircling. Most Harasis campsites in 1980–6 lay to the north of Yalooni (Fig. 8.1). This area has relatively undulating terrain with runoff channels. In the immediate release area the oryx herds ranged south of an approximate east–west line a little north of the enclosure. Their area consisted more of open, stony slopes. Although the project discouraged settlement in the southern sector to allow oryx free access to the release site, the rocky surface deterred *bedu*. Thus, competition around Yalooni for space for oryx and people was scarcely evident.

In 1982 a Royal Order forbade water development in the vicinity of Yalooni. This directive was issued to prevent areas surrounding any new local

Fig. 8.1. Total area used by oryx (shaded), 1982–6, in the release area, and Harasis campsites (●) over the same period. Broken line indicates a road.

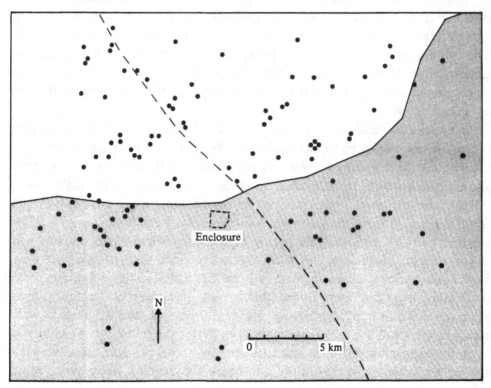

water sources from being so heavily used and settled that the re-introduction area was no longer suitable for oryx. The zone affected by this rule was later interpreted to have a radius 30 km from Yalooni. It was invoked once in 1983 by the Ministry of the Interior to change a prospective water drilling-site 17 km from Yalooni to one 38 km north of Yalooni (Fig. 4.3).

These two regulations related to a fixed area around Yalooni, beyond which the released oryx soon moved. A further two arrangements were made with the Harasis as the need to give the oryx some protection from displacement and disturbance by vehicle movements became apparent (see below). In the summer of 1983, the tribal leaders, the shaikhs, agreed with the proposal that a traditional grazing reserve (4.8.4) should be declared around the first herd in the wild. This was desirable because, following heavy rain in April 1983, the herd ranged widely (6.3.2) before settling for the summer in a large *haylah* with good grazing and shade. The oryx showed every sign of spending their second summer in the desert, independent of drinking water by eating lush vegetation and adjusting their activity patterns (6.9.4). However, the changing distribution of *bedu* camps showed that the oryx area would be heavily settled before the end of the hot months, in which case the herd would move to less good pastures, probably becoming dependent again on water at Yalooni. The Harasis understood the significance of allowing the herd to be water independent as part of their gradual adaptation to the desert, and agreed to an oryx reserve enclosing all the *haylahs* currently used by the herd. The same rules as applied to any other such reserve after rain were agreed upon (4.8.4). The reserve was voluntarily observed from June until August when further, unexpected rain fell which changed the distribution of both oryx and people.

After several incidences of oryx being displaced by Harasis camps (see below), it was suggested to the shaikhs in 1986 that, if the oryx had found an area of adequate grazing or had reached an area of recent rainfall first, they should be allowed to remain there. Specifically, no family would settle within 5 km of the area used by the herd, which would allow the livestock to forage in all directions from the camp without necessarily encountering oryx. The wisdom and desirability of preventing oryx displacement by this means was acknowledged by shaikhs and *bedu*, but compliance and enforcement were impossible. After a local rainstorm in herd 2's current range in January 1986 (6.4), the project rangers found themselves in a dilemma because their relatives and own families moved straight into the area, and the shaikhs were too far away or unavailable to visit the site. Reporting by patrols of what was actually happening on the ground became less realistic, and soon the oryx moved out or returned only at night to drink the standing water and feed on the fresh grass. As the rain area became very densely stocked with goats the oryx no longer returned. At this stage, the Harasis agreed that the oryx certainly should not be displaced but that, if rain was local and offered their stock the chance to graze as a break from supplements, no family could fail to respond by moving in.

8.5 Competition with oryx

The re-establishing oryx share the range with other domestic stock and wildlife, most importantly gazelles. If dietary overlap and shortage of resources led to competition, the re-introduction might fail. Casual observations and discussions with Harasis, indicate that oryx and sheep graze almost exclusively (Table 8.4). All species eat the pods of *Acacia tortilis* and *Prosopis cineraria*, which have very high nutritive values (Table 6.3) but occur only seasonally and with widely varying yields between years. All species eat ephemeral forbs after rain, but the oryx diet is still mostly grass even at this

Table 8.4. *Diet composition of large herbivores on the Jiddat-al-Harasis*

Food	Species				
	Camel	Goat	Sheep	Oryx	Gazelle
Acacia tortilis					
Upper canopy	X				
Lower canopy		X			X
Pods	X	X	X	X	X
A. ehrenbergiana					
Terminal leaves	O	X			X
Prosopis cineraria					
Upper canopy	X				
Lower canopy		X			X
Pods	X	X	X	X	X
Grasses					
Perennial tussock spp.			?	X	
Annual or perennial tufted spp.	X	O	X	X	
Herbs					
Annual spp. after rain	X	X	X	X	X
Woody herbs	X	X	O	O	X
Non-woody perennial spp. with seasonal regrowth	O	X		O	X

X, major diet component
O, minor diet component

time. Many non-woody perennial herbs show some seasonal regrowth from fog moisture (4.5), and the oryx eat small amounts of some soft herbs such as *Convolvulus* sp. These are more important food sources for goats and gazelles. The only woody forb eaten by oryx is the wiry *Crotalaria aegyptiaca*, of which the fruits and leaves are eaten by gazelles, but the foliage is eaten little by camels or goats.

These observations suggest that, although each species could be ranked in a grazer–browser continuum, there is much overlap between diets and, indeed, in a desert area where any green matter is scarce this is not surprising. The overlap of distributions and the relative numbers of each species are more likely indicators of the occurrence or severity of inter-specific competition. Between the grazing species, oryx and sheep, there is little competition because sheep are so scarce on the Jiddah compared to goats. Thus, the widely distributed perennial tussock grasses are almost exclusively an oryx food resource. The diets of oryx and gazelle are very different and the two mingle without apparent effect on one another. Camel and oryx diets overlap more, but camel density is low and they are spread over much larger areas than those currently used by oryx. Moreover, unless browsing on a tree, a camel moves and feeds so that many plants are untouched between successive plants which are bitten. Goats have a heavy impact on vegetation locally, and around *bedu* camps even the tufted grasses, such as *Stipagrostis* sp., are eaten down to ground level. However, oryx are unlikely to occur simultaneously in such areas (see below). Moreover, this very heavy grazing pressure produces a strong regrowth from fog moisture some months after the goats have left, to the oryx's advantage (6.3.7).

Hence, dietary overlap between oryx and goats or camels is only likely to lead to competition if grazing conditions have brought the species into the same area. If oryx vacate the area under these conditions, dietary competition may be irrelevant. Competition for space between oryx and Harasis and their stock occurred under two sets of circumstances, and in both the oryx moved away.

(1) Rain falling over a small area attracted large numbers of people and wildlife. *Bedu* in cars located rain areas within 24 h and reserved campsites close to standing water. They returned as soon as possible, first with empty drums to fill, and then with livestock and belongings. If people arrived first, oryx were unlikely to immigrate into a densely settled area. In contrast, in early February 1986, herd 2 was living 15 km east of Yalooni, exploiting a little vegetation regrowth and being supplemented, when a light shower fell on them. One small pool of water provided drinking water for

7 days. Nineteen days after the rain, one Harasis family moved into the area, which became highly disturbed by vehicles driven by children rounding up camels. At the time a female oryx, her 4-day calf and adult male had separated 3 km from the herd and were living less than 2 km from the camp. Within 24 h the calf had disappeared and was never seen again. Only 7 days later, the 17-day calf of another experienced mother, with a male, living 2 km from the camp and within the area used by its goats, disappeared. From its tracks, it was frightened while lying-out at night and bolted. It was subsequently tracked for 33 km and recovered, severely dehydrated. After a separation of over 72 h it was successfully reunited with its mother, back in the herd, but succumbed to *Salmonella* infection two days later. There was no unequivocal explanation for these two separations, but both took place 2 km apart and within a few days. They were the only such instances of calf loss on the project, and they were the only calves of this age living near a *bedu* camp. While the Harasis were themselves much concerned by the incidents, they did not feel at fault, and these losses were the stimulus for the attempt to limit families from moving to within 5 km of established oryx.

(2) Displacement of oryx by *bedu* settling also occurred under drier conditions. For example, in September 1985, herd 1 extended its range into the western, outer ring of the Lahob meteorite crater (Fig. 4.2), where it discovered a reasonable regrowth of grass in an area of 100 ha heavily used by livestock the previous winter. Apart from occasional returns to Yalooni only to drink, the herd grazed this area until in late October two Harasis camps settled within a kilometre of the herd. On four nights the herd drank water from the goats' supply, but the oryx then left the area to return to the drier, eastern side of the crater. In addition to this incident, there have been instances where project staff have profited from the ability of oryx to detect small areas of superior grazing by moving their homes to these sites.

It is inevitable that oryx will be displaced by arriving bedu households. However, except where there were young calves which were obviously more susceptible to fatal disturbance while lying-out in the first month of life (2.4.4), the oryx's habituation to people was such that human activity at 1 km distance did not alarm them. Observations on oryx and goats coinciding by day reveal a marked antipathy by the oryx towards the goats (and fear by the latter), and the Harasis maintained the two were similarly incompatible in the times of the aboriginal population. As camels and oryx mingle freely, the arrival of goats in an area is probably responsible for oryx departing, thereby lessening the chance of dietary competition but displacing the oryx into less good habitat.

8.6   The Harasis as wildlife rangers

Every male Harsusi had a fund of knowledge and experience of the Jiddah, its livestock and wildlife, and in rare cases, only, had this been eroded through living or working in other areas, or by absence for long periods of education. Any newly recruited oryx ranger knew the terrain, landmarks and areas' names of the Jiddah, could assess the age of animal tracks and follow them across stony surfaces that carried only light and partial prints. He could recognise his own and many others' camels from a distance, and knew his and his relatives' camels from their tracks. These were obviously useful qualifications for working on the oryx re-introduction.

Formal training was confined to radio-tracking practice (5.3.2), so that all rangers could locate and trace a signal very precisely. Use of radio communication equipment encouraged proficiency in sending accurate reports to Yalooni, first over VHF handsets from the enclosure and then over HF radio from a patrol car. These reports usually concerned the size and composition of oryx groups, their activity and location, using distance and direction from known landmarks.

When on an oryx surveillance patrol, the two rangers in a vehicle were controlled from Yalooni by the manager and head-ranger, but were inevitably left to act much upon their own initiative. In the event of vehicle malfunctioning they could usually make a running repair. If there was a radio breakdown or silence, they assessed priority tasks and the most efficient way of covering ground to search for all oryx groups. It was more important to obtain daily locations for groups with females than for single males, which were likely to be in a fairly circumscribed area.

The rangers made accurate observations of animal activity, such as the times of shading (Fig. 6.1), and on the identity of oryx that might be missing from their herd at daybreak. Animals involved in sexual activity were correctly identified. When they knew the oryx well, the Harasis made very fine distinctions between condition or coat colour of different animals in a herd or between herds, usually with an accompanying explanation in terms of grazing or subtle climatic differences.

If oryx were missing, the rangers learnt that they must return to where the missing animals or complete herd was last seen to search for separating tracks.

Although time-consuming and more effort than driving around between places previously used by the missing animals, this technique rarely failed. Once tracks of a group were found, they could be followed over the sandier surfaces in a vehicle at 40 k.p.h. A single set of old tracks over much of the Jiddah's stony pavements might mean seeing one incomplete print almost filled in with drifted sand every 200 m and 20 min. To the outsider, such clues seemed an act of faith. Once tracks were found, the Harasis very rarely failed to find the oryx in due course. Some rangers knew individual oryx tracks, as for their camels.

Although the Jiddah appeared featureless to the outsider, it had abundant landmarks for the Harasis, mostly distinctively shaped *Prosopis* trees, slight outcrops or ridges, or trees which were occupation sites of known people many years ago. This intimate knowledge allowed two patrols to arrange by radio to meet at a specific tree. It also enabled rangers to guide visitors across the Jiddah or to return to the precise site of any incident or sighting after many weeks if necessary.

These activities were highly regarded work. Monitoring of the oryx alternated between bursts of great activity as herds split, rain fell or seasons changed, and oryx moved long distances, and quiet periods when the oryx behaved predictably from week to week. This alternation was very similar to that of a traditional pastoral existence, and appealed to the Harasis. Stand-by duties in camp were found to be more irksome, and provoked many excuses for being somewhere else (8.2.4).

8.7 Attitudes towards the re-introduction project

When the re-introduction project started, there were very few alternative sources of long-term employment on the Jiddat-al-Harasis. The next few years saw the completion of Haima Tribal Administrative Centre, with numerous jobs in ministry local offices, and the development of several oil camps: Marmul in 1981, Rima in 1983, Bahja in 1985 and Nimr in 1986 (Fig. 4.6), with Harasis employed at each. The availability of these alternatives was apparent at Yalooni as the number of Harasis requesting work or applying to fill any vacancy declined sharply. In 1986 the oryx project was no longer the single largest employer of Harasis (Table 4.3), but it still brought over RO.5000 per month into the tribe in salaries, apart from any other benefits. Against the increase in Harasis staff 1980–6 (8.2.1), the only *bedu* to leave were three rangers (one for medical reasons) and one assistant cook (to attend school). This very low turnover suggests that the work was congenial and adequately rewarding.

In general, employment under government terms was favoured over working for a company where there was less job security, although salaries for drivers were twice those on the oryx project. However, working in the desert with oryx and with scope to manage a few camels while in camp under more agreeable living conditions was felt to offset the salary differences to some extent. The project Harasis were drawn nearly equally from the tribal subdivisions, leaving no section disaffected. Their positions were highly valued, and the Harasis took pride in working for the Diwan of Royal Court rather than a ministry (3.7.2). The force of uniformed rangers and drivers had an *ésprit de corps*, although tolerance of any more quasi-military practices or organisation was very low. The oryx project was also well known within Oman through television, firmly associating the tribe with the animal which was regarded as a unique national resource.

The re-introduction project was drawn, to its advantage, into the typical network of mutual obligations which were once probably critical for survival in desert societies (Aronson, 1980; Sandford, 1983). Many Harasis lived too far away to be influenced by the project more than occasionally, but for many it was a last resort source of material assistance, without affecting their lives routinely or making many demands on them. Therefore, the project, which could have been disrupted or even rendered unworkable very easily by a hostile local population, was welcomed. No one said that returning the oryx was a worthless objective, and gratuitous reports of oryx sightings by non-employee Harasis greatly assisted monitoring and indicated a desire to help towards the common end.

The main criticism of the project by local people concerned further development at Yalooni. Haima, the seat of local government and 80 km from Yalooni, was sited there more for reasons of water supply and communications, lying on the Sultanate's main arterial road, than for the convenience of the Harasis (Fig. 4.6). Yalooni would have been geographically more convenient for schooling, medical facilities or a few shops. However, the 1982 ban on water development in the vicinity of the release area aimed to help to maintain its carrying capacity for oryx (8.4). Until the last oryx has been released from Yalooni or the wild oryx no longer have a useful attachment to the release area, the same argument holds for preventing Yalooni from acquiring new functions. The camp was built for the oryx project, and should not become a trading centre with inevitable environmental damage from uncontrolled vehicle access, the building of shops, stores and administrative buildings. In due course, the camp will become the head-quarters of the national nature reserve, in which development will be carefully controlled (IUCN, 1986).

The oryx project was also claimed by some Harasis to be against the development of new graded roads on the Jiddat-al-Harasis. Part of the area's attractions for the re-introduction lay in its remoteness and relative lack of roads (3.7.1). Despite frequent inconvenience, the project and oryx undoubtedly benefited from less disturbance due to the lack of any road into Yalooni. No government intention to grade new roads was ever discouraged, and a properly designed network which would follow the main lines of Harasis movement and the needs of the nature reserve would help everyone.

Even though the Harasis biased their answers towards what a questioner wanted to hear, these criticisms were not frequently heard, and most *bedu* were aware that the interests of the project coincided with theirs in maintaining the productivity of the desert for the benefit of wildlife and livestock. Project management intervened on occasion on behalf of the *bedu*: a seismic party was asked not to drive heavy vehicles through the grazing of the Lahob crater, and a large seismic camp was diverted from settling on the immediate runoff area of a rainwater pool that was to provide water for many families for 3 months. Such actions were appreciated by a large sector of the tribe who were not necessarily employees of the oryx project.

8.8   The implications for oryx of possible changes in Harasis land-use and socio-ecology

With the advent of modern development to the desert regions of Oman over the last 15 years, the Harasis' way of life has undergone profound changes (Chapter 4). A Harsusi employed on the oryx project had many demands upon his time and resources (8.2.3), leading to a very different pattern of land-use than in the pre-motor era. Because the oryx were influenced by human activities, the prospects for re-establishing a viable population of oryx were intimately related to human desert use. The most important factors for the oryx are future patterns of use, the productivity of the Jiddat-al-Harasis and the numbers of Harasis. The trends observed over the duration of the re-introduction project allow some prediction of *bedu* socio-ecology in the short-term future.

A very simple model shows the factors which determine the present settlement pattern of the Harasis *bedu* (Fig. 8.2). The factors are all very broad categories, which are labelled to emphasise their mutul independence. Each, therefore, includes many aspects: for example, 'external economics' encompasses factors such as the cost of animal feed or a new car, or the availability of government funds for major development works. Each direct factor may be influenced in turn by one or more indirect factors.

The model makes two basic points. The first is that in the present motorised pastoralism of the cash economy, the location of grazing is only one factor determining where these pastoralists settle at any time. Following rain, it is pre-emptive (8.2.3), but other factors assume dominance as conditions become drier. Second, the present settlement pattern is highly subject to prevailing economics acting through several routes. This is evident in the Harasis' efforts to minimise costs (8.2.3), in the universal need to earn cash rather than rely on the traditional products of pastoralism, and it is the ultimate motive for the meagre sales of livestock.

At the moment, wage-earning supports pastoralism except under the best ecological conditions. It is, thus, likely that any further adverse economic developments will increase reliance on working for cash by drawing more people into paid jobs at the expense of the pastoralist system. Other factors will act to modify the extent or nature of livestock-raising: many of the boys presently at school in Haima are unlikely to return to work with livestock, and the start of compulsory, primary schooling for girls could cripple small stock management by removing the critical labour of many families.

The pressures on pastoral activities are not independent of the Harasis' aspirations. These invariably centre around greater water availability and provision by government of permanent housing. The scope for further water development on the Jiddat-al-Harasis is unclear because of underlying

Fig. 8.2. Model of direct and indirect factors influencing Harasis settlement pattern.

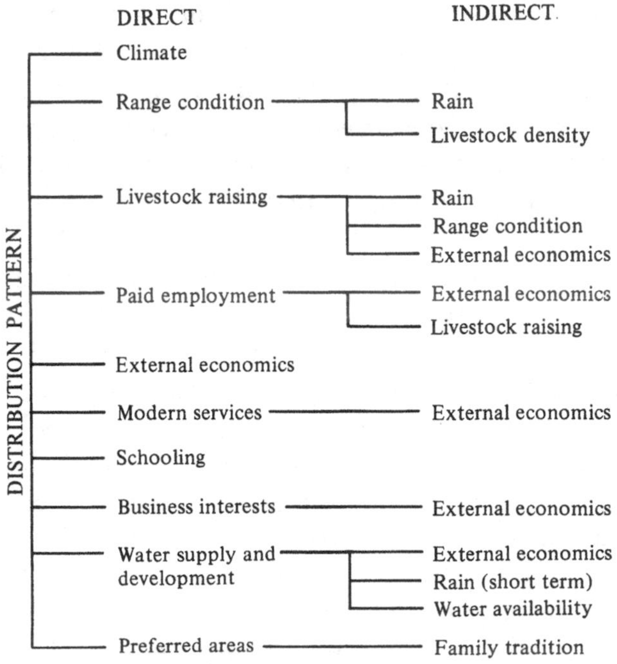

geological anomalies. However, the fact of the main aquifer's recharge area being in south-west Oman (4.3) and the occurrence of potable well water only on the south-west Jiddah (Fig. 4.3) suggest a cline of decreasing water quality west–east across the area. Haima and Al Ajaiz wells probably tap rare pockets of surface runoff, whereas water from other wells on the Jiddah requires expensive treatment to be potable. Thus, further water development will be expensive, and subject to prevailing economic conditions. Providing more water points may serve only to maintain a way of life rather than benefiting the national economy through increased livestock production.

The desire for housing is natural in view of the extremes of winter and summer climate. Although house construction depends on government action, most families are more sedentary each year as the costs, preparations and labour requirement for moving a family, its material possessions and frame dwellings all increase. It is easy to visualise the population becoming semi-settled in all-weather houses, around which minimal- or zero-grazing livestock-raising is practiced. Permanent housing, combined with pastoralism evolving into livestock-rearing often with small-scale cultivation, has been the pattern for nomadic peoples over much of the Middle East's desert fringes in the last 25 years. Motorisation has been the main factor promoting this development (Chatty, 1986). The Harasis may now be on the verge of following this path, adapting stock-rearing, their favoured occupation, to the Jiddat-al-Harasis' particular ecological conditions and its distance from markets and cultivated lands.

The Harasis population is increasing (4.8.3), at an unknown rate. Greater numbers will impinge on all the factors above, but, even if there are more people working for cash, the numbers of livestock cannot increase at the same pace. The feed costs of such an increase would be too high. Thus, the size of individual family members' holdings might decline, leading eventually to some tribesmen owning no stock and relying entirely on their paid employment. There is evidence that men with full-time employment already have 25–35% less stock than those without a salary (Chatty, 1984).

The implications for the oryx population are several. Any tendency for the tribesmen to settle or to concentrate in small settlements will leave large areas open to wildlife. The present dispersion of water sources, both potable and for livestock only, resulted in large lacunae of desert which were not used in 7 years, despite rainfall on some. The oryx with their low water requirement can exploit these areas. The Harasis also agree that there are conditions of grazing which can support a herd of oryx for several months, but which will not provide grazing for a worthwhile period for a flock of goats because of their larger group size and shorter daily ranging.

Unless there are revolutionary and unforeseen changes to the pattern of Harasis life, the oryx are always likely to have living space, provided that the animals can locate the better areas (9.3.1). The creation of the Jiddat-al-Harasis National Nature Reserve will ensure that regional development plans take account of the needs of wildlife. Specifically, the central eastern Jiddah, which has the best tree cover and hence shade (4.5), with the least chance of potable water underneath (4.3), could be a development-free sanctuary, without prejudicing grazing by *bedu* flocks after rain.

8.9   Discussion

As envisaged many years before, the re-introduction project succeeds as a source of long-term employment for desert-dwellers in their homeland. The economic and social benefits from the project permeate the tribe to such an extent that the sense of complete tribal involvement in the oryx re-establishment is strong. The interests of the oryx are also invoked to prevent unnecessary environmental damage during development over a very large area, while the local feeling for wildlife ensures the maintenance of other wildlife populations, such as gazelles and houbara bustard. Such active, community conservation occurs without detriment to the Harasis, as they participate in the development and benefits of modern Oman. This approach to conservation underlies much of the World Conservation Strategy (IUCN, 1980), but pre-dates it.

The natural skill and subsequent experience of the Harasis as wildlife rangers on the Jiddah showed on countless occasions that no one else could follow, locate and monitor oryx groups with comparable skill or motivation. Granted the economic necessity of a salary, employment on the re-introduction project appeals, because of the type and status of the work, to a group of people whom companies regard as notoriously unsatisfactory employees. The cost to the oryx project lies in complex and time-consuming administration, to combine flexible working times with a reasonable amount of work. In this way, home problems and pastoral crises can be coped with, and at the same time oryx protection and monitoring are not neglected. By the same token, flexible management policies mean that the project is undoubtedly exploited on occasion. Staff requests to leave Yalooni camp can be more readily met than in an oil industry camp, and this helps mitigate the discrepancies in salary levels.

Conflict between the project and people occurs only under specific conditions, namely when many occupants and users of the desert are pursuing the same, localised resources. These will almost certainly continue, in future, to be the areas with good grazing after patchy rain. The absence of any security threat to the wild oryx proves the project's acceptability to the local people. Furthermore, the project benefits from a high degree of co-operation from the *bedu*. The Harasis are living under fast-changing economic circumstances, and the oryx have to find their niche in this context. Success will depend partly on the skill of regional planners, and there is a challenge in planning further development so that standards of living rise while the desert's productivity is maintained for all its users.

9

An analysis of project progress, and the future

9.1   Introduction

In previous chapters the performance of oryx re-introduced into Oman's desert, and their wild-born offspring, have been described. Although the number of oryx involved has not been large nor the time-span long, it is clear that captive-bred oryx prepared for release by the methods described in Chapter 5 can thrive in the desert. Chapter 6 showed the main lines of evidence that the oryx exploit their environment in an apparently rational manner. This chapter has two main themes. The first is an analysis of the re-introduction feasibility study and subsequent project design. The costs of the re-introduction on a capital and recurrent basis are then assessed. The second theme is an attempt to model the oryx population with the aim of predicting future performance of the animals. The factors which might help or retard their increase to a viable population are identified, with the management actions necessary to achieve this. These conclusions combine in the final chapter with those aspects of the oryx's biology, environment and re-introduction management methods which aided re-establishment, to attempt a synthesis of re-introduction theory and method.

9.2   Project implementation and cost analysis
9.2.1   *The feasibility study and project design*

The feasibility study (3.7.1) selected Yalooni as the release site for its diversity of vegetation and tree cover, and its location in the central Jiddat-al-Harasis (Fig. 4.2). Other attractions were its adequate proximity to the country's main arterial road, and lack of graded roads in the vicinity, while its reasonable distance from the brackish well of Al Ajaiz avoided the expense of sinking a well (Fig. 4.6). Oryx were recorded at Yalooni in 1964 (Loyd, 1965), and persisted there for some years subsequently. After 7 years of

222

operation at Yalooni, all these factors have clearly assisted the project: the immediate release area met all oryx requirements when grazing was adequate, and even without rain the abundant tussock grasses yielded some green leaf. Most water for the camp is still delivered from Al Ajaiz, which, despite the cost (see below), has the advantage of not encouraging *bedu* to settle in the release area (8.4). The lack of graded roads deterred casual, non-*bedu* visitors from roaming the Jiddah, which has helped to maintain security cover over the oryx area.

The feasibility study drew attention to the discrete area of 7000–10000 km² in the central Jiddah which was better vegetated (Jungius, 1978*b*; see also Fig. 4.2). As the total area used by both released herds lay within this zone (Figs. 6.3 and 6.8), the Yalooni release site was correctly identified as prime oryx habitat. The Harasis agreed. The coincidence of the area used by the re-introduced herds with the locations of oryx sightings in the 1960s (Fig. 2.3) confirmed this. In addition, the feasibility study claimed that the saline flats below the Huqf escarpment never were oryx habitat: apart from a 1-day descent of the escarpment by herd 1 in 1982, no oryx has visited this area. Although the original oryx used the Sahma sand-sea and the western Jiddah seasonally (Fig. 4.2), the failure, to date, of the re-introduced animals to move as far north or west is merely due to lack of inducement. These areas received no rain after the oryx were released.

A central issue of the feasibility stage was whether the re-introduction area had remained adequate habitat through the years since elimination of the original oryx. Obviously, the conclusion was positive, and was based on the evidence that hunting and capture were the causes of decline and extinction. This, with the apparently pristine appearance of the ecosystem (because no trees are lopped for firewood) and the enthusiasm of the Harasis for the re-introduction on cultural and economic grounds, may have masked progressive changes in the habitat and patterns of human exploitation. The feasibility study took place 5–6 years after the oryx's extinction, but this period coincided with the start and rapid spread of vehicle use by the Harasis. This profoundly changed their pastoralism and occupation of the Jiddah (4.8.2). The general effect has been a very large increase in grazing pressure on the Jiddah from an increased human population, which is now free of many of the restraints of water shortage, occupies more of the Jiddah and does so permanently rather than seasonally or after rain. Livestock holdings increased as feed supplements were bought. Some of these changes appeared only in the 1980s, but all stem from the use of cars and the consequent need to earn cash. Their combined effect must have been adverse for re-establishing oryx, either through disturbance directly or by forcing the oryx to occupy areas that were beyond economical reach of water or had suboptimal grazing because livestock were in the better areas. Thus, at the time of the feasibility study, the carrying capacity for oryx might have been lower than in the last years of the original population.

No recommendations were made at the feasibility stage on the project's staff establishment or organisation. The Adviser for Conservation set up the project as described earlier (3.7.2). Good radio communications between Muscat and Yalooni then aided efficient management. Apart from the issue of obtaining further oryx for release, the project's concerns and activities lay entirely within Oman. The simple administrative structure, free from any committee, and the project's institutional position (3.7.2) undoubtedly contributed to progress, and should be considered important in any re-introduction design.

The feasibility report also acknowledged that many aspects of adapting immigrant oryx to Oman's conditions and preparing them for release would have to be left to project staff at the time. However, a breeding herd of one male and two to four females held captive for 5–10 years was envisaged, from which ideal social groups could be established for release at appropriate moments. This approach was not thought practicable (3.7.3), and the enclosure was subsequently built for development of integrated herds (5.2.1). With the exception of this modification, the feasibility study's assessments were perceptive, and project design followed its recommendations closely.

9.2.2   *Cost analysis*

As the feasibility study predicted, the re-introduction would be both long term and expensive. While this continues to be true, the real monetary cost is only relevant in the context of Oman's national economy, which is driven by oil production in modest amounts by regional standards, and by the fact that the oryx re-introduction is a rural development scheme (3.7.1). Construction of the camp and enclosure (3.7.2) in 1979–80 cost approximately 1.25 times the running costs for a year.

In 1986, salaries were by far the largest cost, absorbing 46.5% of the budget (Fig. 9.1). As exactly two-thirds of the salaries were earned by the 78% national staff, the project made a large contribution to the local economy, and was essentially labour intensive. Four other classes of expenditure each accounted for between 8% and 12% of the budget. The largest was vehicle running-costs, due partly to the large distances which vehicles travelled both for camp supply and on oryx patrols. A stony desert is also an environment very hard on cars, through mechanical wear and tyre consumption. Electricity generation absorbed 11% of the budget, and, although demand was very seasonal, there was constant power to run meteorological instruments, radio equipment and deep-freezes, and air-conditioning in summer. The cost of trucking water to Yalooni from Al Ajaiz, or alternatively the Bin Mudhabi reverse osmosis plant (Fig. 4.3) was also very high because of the roughness of the desert and the need for a back-up tanker. Water supply cost almost the same as the entire camp maintenance and administration, including medical and office supplies, radio maintenance, building and machinery repairs, uniforms and consumables. However, as the tanker operator was a Harsusi shaikh, the contract brought further cash to the tribal economy and he gained social credit through his willingness to deliver extra water which was shipped directly to *bedu* households (8.2.1).

Specific research costs, above those absorbed by general oryx monitoring, were low, as were annual provisions for vehicle and equipment replacement. The direct costs of the oryx, even when fed a heavy supplement, were low, and the total project cost would rise little with larger numbers of oryx until any increase in ranger strength became necessary. Many of the project's costs derived from siting and maintaining a modern camp in a hot climate and in a remote area whose natural resources were too scarce to provide building materials or, like water, were distant. The high salary cost would be less with

Fig. 9.1. Apportionment of the 1986 recurrent budget.

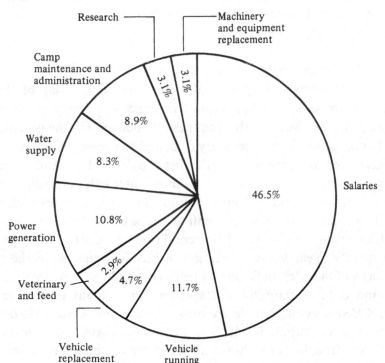

fewer project staff, but the skills of the *bedu* cannot be substituted by any high technology equipment, which, in general, performs unreliably in the desert environment. Thus, the costs of any re-introduction will depend on factors other than the direct costs of supporting the animals, and this may make comparisons between projects invalid. However, the implications of habitat type on re-introduction techniques and costs are discussed further (10.2 and 10.6).

By international standards the oryx project is a very expensive conservation exercise, both in absolute terms and for each oryx. However, this does not imply that all re-introductions need to be expensive, for each depends on its circumstances. Oman's project also has the dual role of providing employment, and for the government it is a highly cost-effective development project. This, and the personal interest of the head of state, ensures that funds are available annually and that support will continue while the project has a role. None the less, the project budget is not exempt from any restrictions on expenditure by the Ministry of Finance.

9.3   The future of the oryx population

The last sections have shown that oryx could be re-introduced on a modest scale. However, can progress be maintained so that a viable, self-sufficient population of oryx roams the desert again, which was the ultimate aim of the re-introduction? The rest of this chapter deals with the very diverse aspects of achieving viability.

9.3.1   *Limitation of the oryx population*

The re-establishing oryx did not face high levels of disease or predation in the desert (7.3.2). Project management was facilitated by not having to protect naive animals from these challenges. On the other hand, the released oryx required supplementary feed, and to a lesser extent water, which might have been a consequence of the diet provided. The resulting relatively constant level of nutrition resulted in short inter-calf intervals, with most births likely to occur over 6–7 months (Fig. 7.2). Without extra feed, breeding would have been lower and conception probably determined, as in most tropical ungulates, by female body condition (Sinclair, 1983), and hence grazing quality. Year-to-year variation in breeding success in the scimitar-horned oryx of the African Sahel was tied to rainfall (Newby, 1974). This, and oryx found dead, apparently of starvation, in drought periods in Arabia (Philby, 1933) suggest that desert oryx populations were limited by food availability. Their population levels probably fluctuated between periods of rainfall and drought, like those of desert small mammals (Newsome &

Corbett, 1975) but over a longer time scale and with rather lesser amplitude. Large numbers of gazelles succumbed on the Jiddah from starvation and/or thirst in the longer drought periods of the 1980s. While the re-establishing oryx remain few in number, management obviously ensures that feeding circumvents these factors and artificially increases the area's carrying-capacity. Starvation is easier to overcome than high challenges from disease pathogens or predators.

Supplementary feeding will be necessary for some or all the population under dry conditions until natural deaths from starvation have no significant impact on the population. While the oryx retain their fidelity to Yalooni, feeding will be feasible. It will become logistically impossible if the animals remain scattered through the desert. Fidelity to the release area is likely to wane when there are adult oryx which never returned to Yalooni with their mothers. The speed at which this stage is reached is unpredictable for it will depend principally on the pattern of rainfall enabling self-sufficiency.

The question of when the presumed natural mortality agent of starvation can be allowed to operate has another side. The oryx's present return to Yalooni for feeding may be a sign that their home-range is not yet large enough to support them through dry periods. As home-range expands only following rain (6.3.2), dry period dependence on Yalooni should decrease after each significant rainfall in new areas. Although the oryx have collectively untilised an area of 3000 km², no oryx have yet discovered the western wadis or the northerly sand-seas of the Jiddah, all of which were good oryx habitat (Figs. 2.3 and 2.6). There is no reason to doubt that when rain falls in these areas the oryx will respond by moving there.

No re-introduced oryx has yet found the natural water sources which were used by the aboriginal population (Fig. 4.3). It is very unlikely that the modern oryx will ever descend the escarpment along the southern Jiddah and cross the now semi-settled coastal plain to drink from the brackish creeks as in the past (4.6.2). There is a greater chance that the seepages along the Huqf escarpment will eventually be discovered, although access to some would be difficult for an oryx. A single, roaming female drank at Bin Mudhabi well in June 1987 (T. H. Tear, personal communication 1987), but neither herd has located this source or Al Ajaiz well, even when living less than 15 km from it. Discovery and use of any source but Yalooni may result in a very different perception of the Jiddah by the oryx. This could then alter their pattern of occupancy and reduce their fidelity to Yalooni. Thus, the oryx still have much to experience and learn, all of which will increase their progress towards self-sufficiency under the worst conditions.

While the oryx need time to master the entire Jiddat-al-Harasis and surrounding habitats, the area itself is subject to change and development. The changing Harasis pastoralism and pattern of land-use have been described (8.8), while the other main impact on the area is the oil industry (Fig. 4.6). The pipeline and oilfields along it occupy a large proportion of the western Jiddah ecological unit, which was known to be prime oryx habitat (see above). The distribution of known but unexploited fields is expanding around Nimr, and north-east of Rima towards the centre of the Jiddah. However, oil extraction need not destroy oryx habitat if new roads and procedures are carefully designed. The earlier stages of seismic exploration are more destructive of the sand surface and vegetation cover. The present pipeline lies buried under a 1 m mound of backfill, which, although visually unattractive, would not be an obstacle to oryx. Disturbance, rather than conflict, is likely to result when the oryx discover the western Jiddah, but, unless the methods of oil extraction change radically, the area will remain potentially available habitat. Declaration of the Jiddat-al-Harasis as a national nature reserve will help to lessen the impact of oil industry operations through the proposed authority's influence on planning and development (IUCN, 1986).

9.3.2   *Modelling the population*

Between 1980 and 1986, Oman's wild oryx population increased by 22% a year through a combination of immigration and reproduction (7.2.1). To estimate the effects of various birth and death schedules and the impact of further immigrants, a spreadsheet demographic model developed by J. Ballou and L. Bingaman of the National Zoo, Washington, was used to make projections. The effects of various alterations were then examined.

The model calculates life table statistics from studbook data, from which basic demographic projections can be made. All projections refer to the size of the female population which can be doubled for an approximate total number of oryx. To obtain estimates of age-specific mortality and fertility rates from an adequately sized sample, information from the immigrant oryx and their Oman-born offspring was augmented by information on the parents of the immigrants. In this way, 75 oryx provided the basic data. The resulting values for age-specific survival and probability of producing a female calf are given in Table 9.1. The actual rate of increase is 108.05% per year. Based on Leslie matrix-type population projections from life tables, 14 wild females in 1986 would increase to 78 females after 20 years.

Many of the age-specific values seemed unrealistic for the wild, and were adjusted (Table 9.1). Survival in each of the first 4 years was decreased, and the probability of producing a female calf at 1–2 years was reduced to zero.

This probability was then held constant at 0.400 between ages 4 and 10, when it was halved until the chance of breeding at 14–15 years was zero. To simulate desert conditions, life expectancy was also shortened so that all oryx died at 15–16 years. These changes reduced the annual rate of increase to 107.9%, and predicted 69 females after 20 years.

In the expectation of further captive-bred oryx going to Oman (9.5.1), a second set of runs of the model retained these adjusted probabilities but, in each of 5 consecutive years starting in 1987, seven female oryx were added to the population. This represented the female component of hypothetical groups of sex ratio 3.7, and the female age composition in each year was: 1 oryx of less than 1 year, three of 1–2 years, one of 2–3 years, and two of 3–4 years. The addition of five such groups would result, 20 years after the first addition, in a female population of 206, while 20 years after the fifth addition there would be 291 females (Table 9.2).

Table 9.1. *Actual and modified values of probability of survival* (P_x) *and of producing a female calf* (M_x) *in each year age class, with the finite rate of increase* (λ) *and female population after 20 years*

	Actual		Modified	
Age class	P_x	M_x	P_x	M_x
0	0.917	0.0	0.750	0.0
1	1.000	0.042	0.900	0.0
2	0.934	0.112	0.900	0.112
3	1.000	0.126	0.900	0.126
4	0.955	0.389	0.955	0.400
5	0.900	0.342	0.900	0.400
6	1.000	0.434	1.000	0.400
7	1.000	0.083	1.000	0.400
8	0.926	0.101	0.926	0.400
9	1.000	0.209	1.000	0.400
10	1.000	0.073	1.000	0.400
11	0.879	0.0	0.879	0.200
12	0.831	0.0	0.831	0.200
13	1.000	0.0	0.500	0.200
14	1.000	0.0	0.500	0.0
15	1.000	0.0	0.0	
to 20	1.000	0.0		
	$\lambda = 108.05$		$\lambda = 107.89$	
Female population after 20 years	78		69	

Table 9.2. *The effects of adding seven females in each of 5 consecutive years on the female population after 20 years from the first addition and after 20 years from each addition*

No. years with additions	Interval years	Female popn	Year	Interval years	Female popn	Year
0	20	69	2006			
1	19	93	2006	20	100	2007
2	18	131	2006	20	150	2008
3	17	162	2006	20	199	2009
4	16	183	2006	20	242	2010
5	15	206	2006	20	291	2011

popn, population.

Table 9.3. *Effects on predicted population of variations in population performance with (1) new immigrants added, and (2) without new immigrants. λ = finite rate of annual increase in population size*

		Results	
Run	Conditions	λ	Female popn after 20 years
1. Immigrants added for 4 years			
XX	Base run: P_x, M_x as in Table 9.1	107.9	242
AA	After 4 annual additions, P_x decreases by 25% in each age	79.1	Extinct
BB	As for AA but P_x decreased for ages 0 and 1 years only	101.9	93
CC	As for BB but M_x increased: year 2 from 0.112 to 0.200 year 3 from 0.126 to 0.300	103.4	120
2. No further immigrants			
YY	Base run: P_x, M_x as in Table 9.1	107.9	69
DD	P_x as in YY, but M_x increased as in CC	110.0	95
EE	M_x as for DD, but P_x decreased by 25% for years 0 and 1	103.5	34

P_x, probability of surviving through a 1 year interval.
M_x, probability of producing a female calf in a year interval.

The results of a further set of runs are shown in Table 9.3. Run A was based on the premise that, once the wild female population reached 50 oryx, management effort might decrease, resulting in a 25% increase in mortality through all ages. This would lead swiftly to extinction. In run B the increased mortality affected only the first two age classes, and the population returned an annual increase of only 102%. When breeding probabilities for years 2–3 and 3–4 were then increased (run C), the rate of increase rose to 103.4%. In the final runs (Table 9.3), no further immigrants were added. When breeding rate was increased as in run C, the population increased at 110.0%. If mortality in the first 2 years was then increased as in run D, the rate of increase declined again to 103.5%.

These simulations suggest that, without any further immigrant oryx, the population will be only 138 animals after 20 years. On the other hand, if 35 new females are gradually added, the population will number over 400 oryx in the same year. As the re-introduction started with only nine immigrant females (Table 5.1), the effect of the 35 is not surprising. The subsequent runs show clearly the perils of allowing natural mortality across all ages to increase as the population builds up, and this may influence project design in future. Improving the breeding rates of 3- and 4-year-old females improves population increase, and the upper values used are probably more realistic. Females born in periods of good grazing can show advanced sexual maturity (7.2.5), and although these times occur unpredictably, they would contribute to the fast increase of which oryx are capable under optimal conditions (3.5.3 and 3.8).

The model shows the population's great sensitivity to juvenile mortality. As most mortality occurs in the first year both in captivity (3.5.2) and in the wild (Table 7.7), management must be directed to promoting calf survival. As early survival is related in Oman to inbreeding level (7.5.3), this model provides overwhelming reasons for augmenting Oman's population with more immigrants to boost population size and increase genetic variability (9.5).

9.4 Minimum viable population size

The most critical question for the oryx re-introduction is the population size at which the oryx can be assumed to be self-supporting and secure. While absolute figures are hard to estimate, there are two aspects to this assessment, namely the suitability and availability of habitat, and the population's characteristics, especially its demographic and genetic structures.

9.4.1   *Density and carrying capacity*

The area of potential oryx habitat continuous with the release area extends over much of Arabia. However, the Jiddat-al-Harasis ecological unit and adjacent sand-dune area covers 34000 km² (4.2), of which an insignificant proportion is lost permanently to other uses. The actual area available to the animals depends primarily on the extent of adequate grazing, which is then modified by the numbers and distribution of people and livestock, and the location of water points.

Erratic rainfall causes the desert's carrying capacity to fluctuate widely. On the basis of significant rainfall at Yalooni in 1977, 1982 and 1983 and none subsequently, any part of the Jiddah might receive rain once every 5 years. Ignoring any carry-over effect of rainfall on grazing into subsequent years and growth due to fog moisture (4.5), the oryx are left with 6800 km² of fresh grazing per annum from the entire 34000 km². Under present conditions of sharing their range with livestock, and at an average density of 22 km² per oryx (6.3.3), this somewhat arbitrarily defined ecosystem would support about 310 oryx.

9.4.2   *The demographic minimum viable population*

Conservation biology is acutely concerned with the theory of minimum viable populations in the wild (Gilpin & Soulé, 1986; Soulé, 1986). This concern is derived partly from zoos' intense efforts to sustain small captive populations through sound genetic management (Ralls & Ballou, 1982). It yielded the first estimation of the minimum effective population size (N_e) of 50–500 individuals, which would prevent an immediate decline in population fitness due to inbreeding depression (Franklin, 1980). Inbreeding in the re-establishing oryx was discussed earlier (7.5). At the level of wild populations, attention is now focussed on extinction as a systems process (Wilcox, 1986). All populations have some probability of extinction through the action and interaction of a range of deterministic and stochastic factors. Small populations are particularly susceptible to stochastic factors, such as sex ratio, genetic drift and inbreeding, which are functions of internal population processes. Despite the gloomy terminology, these models are as directly relevant to small but establishing populations, such as the oryx, as to truly wild but diminishing populations.

In a recent conceptual model, population vulnerability is analysed in terms of three vortices into which populations are sucked to extinction: population phenotype, the environment, and population structure and fitness (Gilpin & Soulé, 1986). The causes may be deterministic factors or stochastic events, which may act singly or with positive feedbacks to one another or to the three fields. The action of the model's extinction factors (Table 9.4) has already been seen in the oryx, or can easily be imagined. The run of ten male calves sired by a single male (7.2.4) or dominant male infertility (7.4.2) are such stochastic extinction causes. The demographic models (9.3.2) were deterministic in their predictions of population extinction without the intervention of further immigrants. To persist, the oryx must overcome this range of interdependent and re-inforcing effects (Gilpin & Soulé, 1986). The only extinction vortex to which the oryx are probably not susceptible is fragmentation of their desert habitat. Symptoms of the other environmental vortices have already been seen. The consequences of genetic drift, due to a small number of founders or restricted genetic variation, may only be apparent in future generations, emphasising that the ultimate cause of extinction may precede the final event by many generations (Wilcox, 1986).

The model may be particularly appropriate to the oryx: first, it predicts that large species, bearing single young, may persist at low population levels because of individual longevity (3.5.1) and physiological buffering from short-term environmental changes, but such species are more likely to be sucked

Table 9.4. *Causes of population extinction, from Gilpin & Soulé (1986), and their observed or possible effects in the oryx population. N is population size, N_e the effective population size*

Model Extinction factor	Oryx Observed or possible effects
A. Deterministic	Greater human numbers occupying larger area of oryx habitat Loss of grazing through increased livestock
B. Stochastic 1. Demographic	Small population size Infertility of dominant male Male-biased sex ratio
2. Environmental Chance lowering of population size and increase in variance of population growth rate	Period of rainless years, causing reduction in breeding rate
Fragmentation of habitat	Not applicable in desert
Inbreeding depression due to inbreeding, reducing N_e/N ratio	Calf survival related to inbreeding level N_e/N ratio low because of herd dynamics and sex ratio
Consequences of genetic drift due to small N_e	Longer-term failure of population phenotype to fit environment

into the vortices of inbreeding depression and genetic drift. Second, compared to individuals in natural but diminishing wild populations, the re-introduced oryx are likely to be less fitted genetically to the local environment because the founders were a random selection from those available (7.5.1). This would explain the mortality of some oryx which were evidently incapable of adaptation to the desert although not inbred (7.5.3).

Minimum viable population theory cannot prescribe a single 'magical' figure for all populations under all circumstances (Gilpin & Soulé, 1986). An actual wild population of over 200 oryx could contain an adequate effective population size for genetic considerations, and certainly many populations of Arabian oryx must have lived and bred at lower densities than one in 22 km². However, extinction ultimately through loss of genetic diversity must have been a possibility even without direct persecution. Moreover, in many natural populations, local extinctions could have been succeeded by re-establishing immigrants from neighbouring areas. As this is not currently an option for Oman's oryx, it is reason and justification for continuing management intervention more intensively than would be necessary for an established, natural population. The lessons for the re-introduced oryx are clear, namely that the chances of the population persisting depend on widening the genetic base to reverse the contraction which has already occurred in Oman, with its adverse effects of rising inbreeding level (7.6).

9.5   Importation and release of further oryx

Various lines of evidence have shown the importance of increasing the number of immigrant oryx in Oman, both to accelerate establishment of a viable population and to reduce the probability of stochastic factors causing a second extinction. Between the decision or agreement to supply more oryx and their actual genetic contribution to the wild population, there are many complications which are relevant to any re-introduction of captive-bred animals. These divide into the selection of animals and their subsequent management in Oman.

9.5.1   *Selection of further immigrants*

When selecting future immigrants, there are two obvious, albeit slightly contradictory, criteria. Although the World Herd founders were fairly evenly represented in Oman's 1985 population (Fig. 7.6), the living descendants of those under-represented would be obvious candidates for Oman. This would reduce inbreeding build-up. On the other hand, as some of these animals would not be expected to fare well in the desert, as with the founders, a prudent alternative would be to select oryx related to individuals which had already proved themselves in the desert. Individuals have been identified objectively on the first criterion (Mace, 1986). For maximum genetic diversity some oryx should also be sought from the better-managed Middle East herds (3.10). However, the advantage of their unrelatedness to World Herd oryx would be countered by ignorance of their pedigrees or of their performance. As Oman's present population is almost entirely derived from US stock, its future herds should be assembled with breeding males from the Middle East and females from both populations.

Oman had realised the need for further immigrants for several years, but, following the 1980 dissolution of the World Herd (3.9), there was no body with an overall interest or responsibility for oryx management which could make animals available. The claimed high commerical value of Arabian oryx, and the fact that the US population, while still expanding, is now held in small groups by many institutions (3.5.4; Mace, 1986) may have contributed to the lack of animals for re-introduction. As the captive population seemed secure, there may also have been complacency over the accuracy and completeness of studbook recording. This is most regrettable in view of the value of pedigree data to analysis of the re-introduction.

The lack of a unified management authority was rectified in 1987 by the American Association of Zoological Parks and Aquaria establishing a Species Survival Plan (AAZPA, 1987). Its objectives are to maintain a demo-graphically stable captive population of 350 oryx in the USA, with genetic diversity through equal founder representation, and to meet an annual quota of 10 oryx for re-introduction in each of the next 10 years. Starting with eight institutions holding 190 oryx (Homan, 1988), its effectiveness will depend on the extent to which all captive herds belong to the scheme and are managed co-operatively. This will, in turn, raise problems of ownership and disposal (Conway *et al.*, 1984). A second Species Survival Plan for the oryx in European herds was also created (G. M. Mace, personal communication 1987). Even with these plans and schedules, there remain the previous formidable problems of assembling animals for shipment from different zoos, and of processing export applications. International transport of animals is beset with increasingly restrictive veterinary regulations. These restrictions are a spur to research in reproductive technology (Hearn, 1986), and zoos now require fewer wild-caught animals (Knowles, 1986). However, animals for re-introduction must themselves be transported, and the day may come when they must originate in the country or region of eventual release.

In addition to genetic criteria, potential immigrants must exhibit no behavioural abnormalities, even if captive individuals cannot display the full range of natural behaviour. The two oryx unfit for release in Oman had been subjected to highly unnatural conditions of confinement for the sake of display, in one case at a critical age and in the other for a prolonged period (5.2.5). The maintenance of wild-type behaviour in the face of captive selective pressures over many generations is a significant concern in re-introduction prospects (Frankham *et al.*, 1986).

9.5.2   *Adding new immigrants to the wild population*

Strange animals cannot be added to established herds in the desert without a strong chance of rejection, leading to death in desert-naive animals. Even if new animals were tolerated, their viability amongst animals more adapted physiologically to the desert's rigours would be lower. For both reasons, now immigrants must be released as complete herds, as already done in Oman. Future management should follow the same, proven principles (Chapter 5), ensuring that the first summer is spent in captivity for acclimatisation. The start of breeding will be encouraged by not transporting females around the age of sexual maturity (7.2.8). This will influence the optimum combination of age and sex for batches for new oryx. The total time in captivity can be reduced if the herd members arrive in Oman together or over a short time, and because the relative value of each oryx decreases with overall population increase. New immigrants should be protected by vaccination from catastrophic loss through rinderpest, and covered against botulism for as long as possible because of its known threat (7.3.2).

These second-wave herds would be genetically different from the wild population, and would remain distinct until animals exchanged. The period between release and such meeting and mixing cannot be predicted, being dependent on factors such as the distribution of the two sets of oryx. However, experience shows significant exchange rates may take several years (6.5). Thus, genetic enrichment of the population by adding further unrelated immigrants is not a quick process. The fragile state of the population suggests that new animals should be added as soon as possible. This is predicated on other grounds, namely that the longer the headstart the first herds have in their progressive, genetic adaptation to the new environment the less likely it is that immigrant males will be competitively superior. Thus, the chance of new genes entering the population declines with the period between the two sets of immigrants.

9.6   Discussion

Oman's oryx project shows that on a modest scale re-introduced animals can survive in the desert. Full independence, when natural mortality factors are allowed to operate without management intervention, will take a long time. This is principally because of irregularity in the rainfall, which stimulates range extension, and the resulting knowledge of a larger area then promotes survival during future drought periods. The vagaries of the desert climate cause alternating periods of total independence in the oryx, followed by partial dependence again when they return to Yalooni. The rate at which re-introduced animals become independent and the way by which this is achieved will vary with circumstances and the species, but clearly it does not need to be a continuous and uninterrupted process starting at release.

The oryx in Oman have accumulated a large fund of knowledge and experience of their ancestral environment. This varies between individuals, depending on their original herd and its movements, inclusion in splinter groups, the different inclination of individuals to wander or explore and the various reasons for doing so, and so on. This diversity of experience and perception of the environment, and in the pattern of its exploitation, has developed in less than 5 years in the wild for the first released oryx. The documented observations behind these deductions argue convincingly that captive-bred oryx respond to their natural environment, adapt their behaviour and accumulate information from their experiences. This information is stored and recalled to advantage after long periods of time. As this process is continuing through the life-spans of oryx which arrived in Oman as adults, the lifetime experience of animals born in the wild should be considerably greater. It is this expectation that diverts so much effort to protecting the founders, without whose survival there would be no wild-born individuals.

The re-establishing oryx's use of their environment shows that a body of experience has accumulated, which is handed on to successive generations. Achieving this stage means that the collective knowledge of the last aboriginal population is being replaced by newly evolved social experience. This cultural transmission of behaviour is particularly valuable in allowing rapid responses to changing environments. But there is little reason to think that the old and new traditional knowledges are the same. It was shown above that the re-introduced oryx have still to reach several areas of the Jiddat-al-Harasis which were formerly good habitat. Even when these are discovered and used, the oryx may develop a pattern of movement and use of the Jiddah which is different from that before the 1972 extermination. This pattern will be determined by four main factors.

(1) The behaviour of the establishing population is influenced by its fidelity to the Yalooni release area. This has overriding advantages to project management, but there would have been no comparable attachment to a single area by all the original wild population. This fidelity will gradually disappear (9.3.1).

(2) Animals' prior experience may influence their lifetime perception of

the Jiddah. Animals released or born in the central Jiddah, the present total range, may retain a stronger attachment to it than areas found only later in life.

(3) The oryx's environment in Oman underwent considerable changes between extermination and re-introduction (4.8.2 and 9.3.1). Therefore, the strategies of the original population might no longer be adaptive today. The re-introduced oryx are necessarily learning to live under the prevailing conditions, both ecological and socio-economic, and while they may never discover certain former water sources (9.3.1), they will benefit and adjust their distribution patterns after finding modern water sources on the Jiddah.

(4) For many generations, and perhaps permanently, the re-establishing oryx will be genetically different from the original animals. Even when the degree of 'fit' between the new population and its environment is high, this may be due to co-adapted gene complexes (Templeton, 1986) which are no less adaptive but, at the same time, different from those of their predecessors. This might result in alternative patterns of habitat use. The possibility that different gene complexes are selected for in the re-established population is reason for maximising genetic variation in the immigrants and population.

The conclusion must be that a re-introduced oryx population is unlikely to show the original pattern of utilisation of its environment, even if this were known. There may be more than one set of strategies to exploit a single environment, and the chances of an environment not changing during the period between extinction and restoration must be slight. Indeed, logically, the removal of a species must result in changes to the ecosystem. In this case, re-introduced populations will be genetic, behavioural or ecological clones of the original one only under exceptional conditions. Oman's new oryx may be visually and aesthetically identical with the original animals, respond to short-term changes of climate and grazing in the same way, but may exploit the changed ecosystem with different strategies and finally settle at a different density. The oryx re-introduction not only has provided a wealth of observations but has generated many speculative predictions. Their value and applicability to other re-introductions are assessed in Chapter 10.

10

Towards a rationale for re-introductions

10.1   Introduction

The planning and execution of a re-introduction is complex, as Oman's oryx project has shown. The main concerns can be classified into biological, non-biological and an interface between the two to achieve re-introduction success (Fig. 10.1). Although developed from experience with the oryx, the model is proposed as applicable to all re-introductions. The arrows are one way, with the exception of a dialogue between the re-introducing agency and the sources of founder animals (3.7.3). Various types of relationship are denoted, with cause and effect prevailing on the biological side, and influence and sequence of action more evident in the non-biological and interface zones. Together, they all converge on, and influence, re-introduction success. The non-biological aspects have already been discussed in connection with the oryx re-introduction (Chapter 3). The primary concern of this final chapter is the concept of species' 're-introducibility' (Fig. 10.1), to see whether any general precepts show which species are most capable of successful re-establishment, should the need arise. The role of captive-breeding policies is also examined.

10.2   Two contrasting re-introductions

Many aspects of the Arabian oryx's biology, behaviour and habitat favoured the animals' re-establishment. These have been pointed out in earlier chapters, but they become more apparent by comparison with a species of highly contrasting habits and habitat. For illustrative purposes the orang-utan, which lives in tropical rainforest, is used (Table 10.1). This table compares, for each species, aspects of their biology, habitats and management methods relevant to their re-introduction.

Data for the orang-utan are taken from a variety of sources which provide a consistent picture of the species for present purposes (MacKinnon, 1974; Rijksen, 1978; Galdikas, 1985; Rodman & Mitani, 1986). Re-introduction of orang-utans has been widely attempted, using mostly orphaned or confiscated animals. Most have been released to re-inforce wild populations. Ecological rehabilitation seems more successful than social integration (Aveling & Mitchell, 1980). Successful rehabilitation depends on each animal's history, with an early period in the wild a great asset, and the skill with which they are weaned from human association and structures into natural habitat. Normal

Fig. 10.1. Factors relevant to a successful re-introduction.

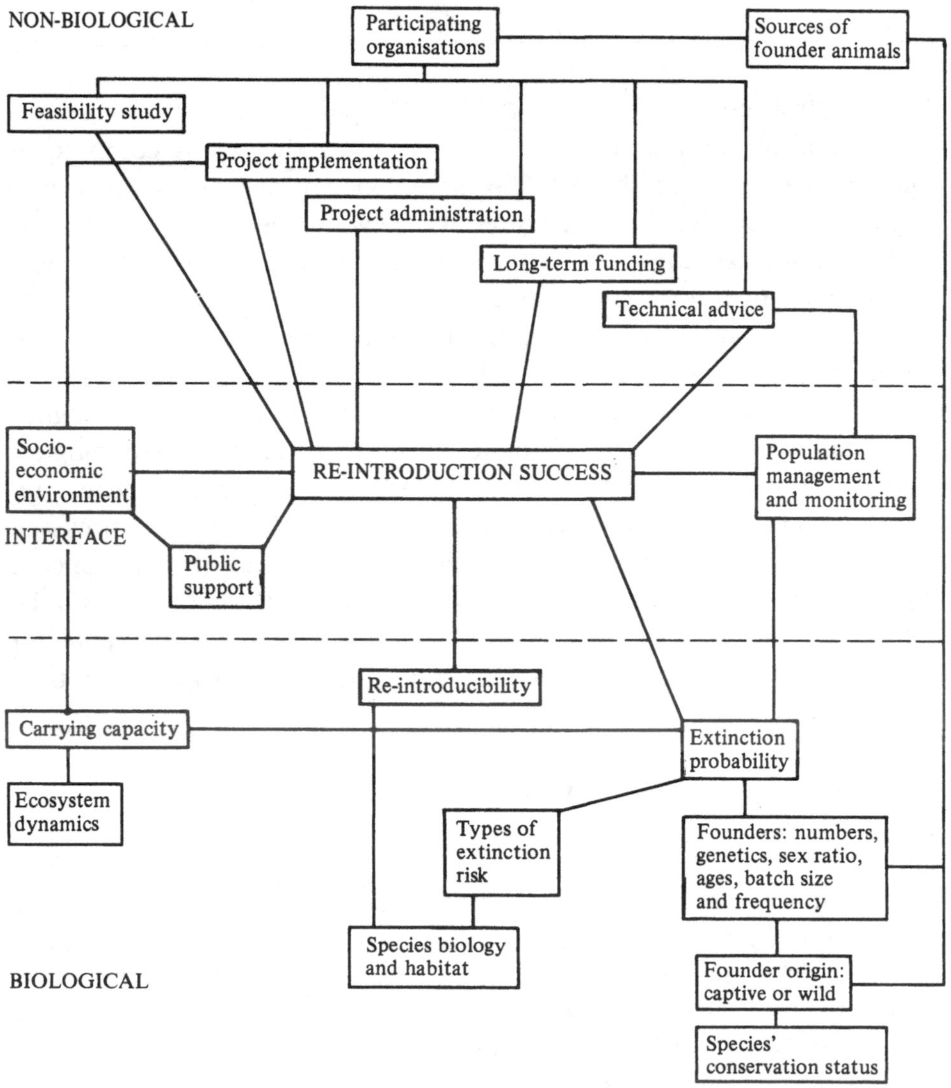

Table 10.1. *Factors relevant to re-introduction compared for two animal species and habitat types*

	Oryx	Orang-utan
A. Habitats		
1. Habitat	Desert	Tropical rainforest
2. Habitat dynamics	Stable stage	Patchy, successional stages
3. Structural complexity	Low	High
4. Productivity	Low	High
5. Predictability	Low	High
B. Animals		
1. Social organisation	Gregarious, fixed membership herds	Mostly solitary, temporary associations
2. Diet	Grass, some forbs	Fruit, leaves, shoots, insects, eggs
3. Diet diversity	Low	High
4. Diet selectivity	Low	Very high
5. Diet seasonal variation	Low	High
6. Learned component of diet selection	Low	Very high
7. Food plant distribution	Low density over large areas	Scattered but clumped
8. Plant secondary compounds	Rare	Common
9. Community diversity	Low	Very high
10. Inter-specific competition	Low	High
11. Risks	Starvation	Disease, predation, starvation, injury
12. Released animals	Groups	Single
13. Range development	Explorer	Conservative
14. Reproductive rate	High	Low
C. Management methods for re-introduction		
1. Visibility in habitat	High	Low
2. Human passage in habitat	Easy	Difficult
3. Animal conspicuousness	High	Low
4. Staff:animal ratio	Low	High
5. Use of vehicles	High	Low
6. Area to be monitored	Very large	Large
7. Security		
No. groups and size	Few, large	Many, small
Impact on population of elimination of group	High	Low
8. Staff time per animal before release	Low	High

behaviour in rehabilitants is promoted by association with more independent or wild animals.

Despite the release of almost 100 orang-utans in Sumatra over 5 years, their success in the forest has been hard to assess. Rehabilitants developed a wild-type diet within a few months, although the process remained obscure. High predation rates were a problem, probably because many orang-utans remained semi-terrestrial unlike the totally arboreal wild animals. Dispersal from the release site was confined to a surrounding area of only 30 ha, which was too small for total independence. Agonistic interactions between wild orang-utans and adolescent or older rehabilitants re-inforced the latter's reluctance to disperse.

The habitats of oryx and orang-utan have many differences relevant to re-introductions. The desert is physically safe (7.3.2) and two-dimensional, whereas the tallest rainforest trees have crowns 60 m high. As with the golden lion tamarin (1.3.1), the forest's structural complexity discouraged normal locomotion in some re-introduced orang-utans, The desert has vast areas of a uniform successional stage, while the rainforest is a complex, mixed forest, with patches at each stage of the regeneration sequence.

The climates of the two habitats are fundamentally different. The desert's erratic rainfall results in runs of rainless years, so that plant productivity is low and unpredictable in occurrence. The forest of North Sumatra is humid, with almost daily rain and a yearly precipitation of 3000 mm. This contrast is reflected in vegetation structure and diversity. While the desert has three species of low trees and 30–40 common grasses, forbs and shrubs, the southeast Asian forest is one of the world's most diverse ecosystems. Not only are there many species, for example 111 tree species in a 4.5 ha plot, but individuals of a single species are usually far apart.

Such habitat differences show in the diet selection and diversity of the oryx and orang-utan. On release the oryx ate almost every grass available. Only one grass was noticeably aromatic and spurned, and one forb was implicated in terminating pregnancy (7.3.2). Both after rain and in dry periods the oryx ate the same staple grasses, augmented by ephemeral forbs. In contrast, the orang-utan's diet comprises fruits, leaves, bark and shoots, insects and miscellaneous items such as birds' eggs and fungi. Plant species and their parts eaten are selected against a high incidence of secondary compounds to deter herbivores. The orang-utan's food sources are dispersed: of 28 chief fruit species eaten in Borneo, 18 had densities of less than two trees per hectare, and none was abundant. In addition, each fruit species is available for only a very short period. Thus, a rehabilitant orang-utan has to develop a varied diet through experience in a forest with complex and irregular fruiting patterns. Although wild orang-utans demonstrate a remarkable knowledge of the most productive parts of their ranges at each time of the year, a rehabilitated animal, reluctant to move far from its release site, faces a daunting task in becoming self-sufficient.

It is evident that an orang-utan can learn to support itself most easily by mixing with wild conspecifics. The social structures of orang-utans and oryx are relevant here. Oryx have a single social unit, the mixed male–female herd with long-term bonds between individuals. In contrast, the orang-utan is predominantly semi-solitary, occurring as adult males or adult females with dependent offspring, or as independent immatures. For re-introduction, the orang-utan's minimal social organisation does not require prior development of a wild-type group, as in the oryx or chimpanzee (1.3.2; Brewer, 1978). This may explain the partial success of rehabilitating orang-utans compared to chimpanzees (Galdikas, 1985). Paradoxically, however, total ecological rehabilitation of orang-utans probably requires mixing with wild conspecifics, but re-inforcing natural populations is currently out of favour because of the risks of disease importation (Rijksen, 1978).

Re-introduction prospects are influenced by community structure. The desert's community of large herbivores, both wild and domestic species, resulted in little dietary competition for the oryx (8.5). The rainforest's herbivore community is one of the most diverse on the earth. Wild orang-utans face dietary competition from a variety of birds and mammals, including five other species of primate in Sumatra (Rijksen, 1978). Despite adaptations to mitigate inter-specific competition, its occurrence is a greater hazard for a re-establishing orang-utan than an oryx. The orang-utan faces further challenges to survival. Large felids are known to take orang-utans, but man has been a significant predator over a very long period. His predation may have encouraged the arboreal life-style, which exposes the heavy orang-utan to injury or death from falling. Rehabilitants often fall. Orang-utans are also subject to a wide variety of internal and external parasites and infectious diseases. While not necessarily fatal themselves, they reduce the animal's ability to cope with other challenges. Thus, the orang-utan's exposure to predation, severe food competition, physical injury and disease is, again, in contrast to observations on the re-establishing oryx. For the latter, the desert's structure, climate and animal community present only starvation as a major challenge.

The success of re-introduction depends partly on the rate at which the species can increase through reproduction. Despite the similar weights of Arabian oryx (55–75 kg) and orang-utan (30–80 kg), and gestation periods of 266 days in oryx and 245 days in captive orang-utans, their breeding biologies differ. A wild female oryx has her first calf before 3 years old, and can then have a calf every 9–12 months. This allows the 22 % annual increase observed in Oman (7.2.1). By comparison, a wild orang-utan may conceive for the first time at 13–15 years (Rijksen, 1986). The inter-birth interval in the wild of 6–7 years is probably the longest of any primate species. The species' low rate of increase would probably prevent establishment of a population within a realistic period without releasing massive numbers of animals.

As the oryx project in Oman showed, re-introduction success depends partially on the ability to monitor release animals effectively. Monitoring in the desert and forest requires different techniques (Table 10.1). Visibility in the desert is good, it is an easy habitat for man to travel through, and an oryx herd is conspicuous. Locating a particular herd of oryx daily is aided by their habit of moving to good grazing and then staying in a relatively circumscribed area for weeks or months (6.3.1). The forest habitat is the opposite in every respect. Locating the same orang-utan on consecutive days is almost impossible without continuous surveillance, because its food distribution demands constant movement.

Although the area used by a group of oryx may be several times larger than that of the same number of orang-utans, vehicles can be used efficiently with a lower staff:animal ratio. The gregariousness of the oryx also results in more oryx being seen per unit effort of searching. Because herds of oryx are released, the cost or effort per released animal is less than for the orang-utan, which requires intensive and personal rehabilitation. These points all suggest that preparing orang-utans for release and then monitoring them is relatively labour-intensive and expensive. The orang-utan has one distinct advantage in that loss of an entire social unit is less catastrophic for the population than for the oryx. This emphasises the need for a benevolent human population in the oryx re-introduction area (Fig. 10.1).

10.3   Re-introducible species

The points made above suggest that the desert and the oryx's adaptations to it were particularly suited for incorporation into the re-introduction methods, and facilitated re-establishment of the herds. The oryx share some of these key features with other animals, and other species have attributes which also predispose them to successful re-introduction (Table 10.2). On the basis of current experience, these appear to be the most re-introducible species. The categories are not mutually exclusive, and the oryx falls into several.

The first class contains generalist species of extreme environments. This obviously includes the oryx, but the muskox, occupying the seasonally frozen sub-Arctic, is in the same category, and has been successfully re-introduced into many sites (see, for example, Smith, 1984). The second group comprises species that can tolerate habitat change. Joining the oryx again, is the European bison. This was re-introduced into Polish forests after surviving solely in captivity, where it thrived because, although a highly specialised animal, it exhibited great adaptability to the changed conditions forced on it (Krysiak, 1967). The habitat available to re-introduced bison had undergone a metamorphosis from primaeval forest through human activities. Also in this category are species which are tolerant of a wide range of habitat conditions. This is probably behind the success in translocating and establishing the black rhinoceros in many places (A. J. Hall-Martin & A. K. K. Hillman-Smith, personal communication). The white and Indian rhinoceroses have both been translocated to set up new populations often in areas far from their sources (see, for example, Booth, Jones & Morris, 1984). The rhinoceros shares the orang-utan's sole characteristic favouring re-introduction, its solitary nature. This means single animals can be released.

Ungulates that move in large, cohesive groups have been introduced or re-introduced successfully, such as reindeer/caribou (1.4) and the European bison. In the cases of the muskox and Arabian oryx, group composition

Table 10.2. *Classes of animals which are most re-introducible*

Class	Examples
1. Generalists of extreme environments	Oryx, muskox
2. Species tolerant of habitat change or of wide range of habitat conditions	Oryx, bison, rhinoceros
3. Species with cohesive groups	Oryx, muskox, bison, reindeer/caribou
4. Large animals	Oryx, reindeer/caribou, bison, rhinoceros
5. Explorer species	Oryx, vultures, condors
6. Scavengers	Sea eagle, griffon vulture, red wolf, condors
7. Species with sanctuaries in habitat	Otter, beaver, hutia, red wolf
8. Nocturnal species	Eagle owl, otter
9. Species whose behaviour can be manipulated	Birds of prey, solitary predators

remains stable over long or very long periods. These species are all at the upper end of the ungulate size range. Large African antelopes tend to run in larger herds, and have simpler feeding habits, taking coarser diets (Jarman, 1974), thus requiring less adaptation by re-establishing individuals. These large animals occupy open habitats, thus aiding monitoring. It is also no coincidence that reindeer (Herbert, 1976), muskox (Wilkinson & Teal, 1984), and beisa oryx (Stanley Price, 1985*b*) have all been domesticated at times, and no doubt the European bison could also be domesticated. The gemsbok has established itself as a successful exotic in New Mexico (Saiz, 1975).

The next category of re-introducible species comprises explorers. The oryx was in this class because the desert's patchy pattern of prime grazing caused phased but continuous expansion of home-range. It was proposed that this process would result in knowledge of a range that was large enough in drought periods (6.12). Re-introduced birds such as the griffon vulture (Parc National des Cévennes, 1981), the bearded vulture (Anderegg *et al.*, 1983), the Andean condor and, probably, the Californian condor, that soar very large distances in search of carrion are also likely to be explorers. These birds are scavengers, as is the re-introduced sea eagle, whose diet is mainly carrion (Love, 1983). The red wolf, due for re-establishment, takes a wide range of small prey, but seems more of a scavenger and opportunist predator than other wolves (Parker, 1986). Feeding on carrion requires less sophisticated techniques than hunting live prey, and offers scope for a wider and more easily found diet. These features would all make an independent existence after release easier for partial or total scavengers than for predators.

Successful re-introductions have been recorded in species whose habitats contain a secure sanctuary against interference by humans or natural predators. The beaver builds its own lodges, and the hutia was returned to areas of Jamaica after artificial warrens were built in suitable substrate, in imitation of other colonies' refuges on the island (1.3.3). Released otters in England were safe during the day by lying up in impenetrable thickets (Jefferies *et al.*, 1986), and the red wolf release area has been selected for its thick cover for sanctuary (1.2.2). The final class is related, and comprises animals that are nocturnal. Although presenting problems for monitoring, animals such as the re-introduced otter and European eagle owl were able to be active at night without disturbance. In the latter case, being nocturnal was probably essential for successful re-establishment in areas of high human density (von Frankenberg *et al.*, 1983).

The final class of re-introducible species differs slightly, for it comprises those for which management methods can instil a long-term release site attachment. While commonly used with individuals, and probably impractical for large numbers of re-introduced animals, it enables a very gradual independence if necessary. It is the basis of hacking in the peregrine falcon (1.3.5), and could probably be used for all birds of prey. Returning to be fed has been used with rehabilitated apes (Brewer, 1978; Rijksen, 1978) and with large carnivores which hunt singly, such as cheetah (Adamson, 1969) and tigers (Singh, 1984). Its value in group-hunting lions with the object of re-inforcing a native population has been equivocal (Adamson, 1986).

10.4   Less re-introducible species

Species for which re-establishment might be particularly difficult include those with characteristics opposite to those of the last section. The attributes of the orang-utan described above (10.2) place it firmly in the less introducible class. It is essentially a highly specialised species, with low reproductive rate, occupying a hazardous and competitive environment, which allows only intermittent and short-term observation of released animals.

Other re-introduction attempts confirm the problems with specialist species. The Hawaiian goose has been returned to its native islands, but despite large numbers of releases over many years, the success with which the remnant wild population was re-inforced is not clear (1.2.1). Part of this may have been due to the birds' unavoidable release into a high altitude zone which had been used only seasonally by the original birds (1.2.2). Apart from this, the goose has several distinctive, specialised features. The adults moult when leading their goslings, which resulted in large hunting offtakes in the past. Both parents and young are very susceptible to introduced predators, of which several species are common (Berger, 1978). The species is unique in its group by breeding as daylength shortens in autumn. As a winter breeder, a higher altitude nesting zone in the re-introduced population might lead to far lower reproductive success. Despite the long interest in the bird, there is little information on nesting behaviour, because released birds have never been followed up (Kear & Berger, 1980). The lack of monitoring has been partly due to the birds occupying lava flows in which man can barely move for the treacherous conditions underfoot. A further sign of its specialisation is its loss of webbing between the toes, adapting it to an almost terrestrial existence far from water (Berger, 1978). Although captive-breeding of the goose was an early and spectacular success, its return to Hawaii has been less impressive for a variety of reasons. Its ecological requirements may still not be totally known, but the species' intrinsic specialisation may be a root problem.

Life history characteristics have been implicated in other re-introduction difficulties. Re-establishment of the great bustard into southern England has made little progress. Although doubts have been expressed about the suitability of the founding stock, originating from areas with very different climates (1.2.4), problems with semi-captive breeding may have been due to the bird being highly specialised with a long life-span and low reproductive output (Osborne, 1985). The pilot release of captive-bred golden lion tamarins in Brazil was into their natural forest habitat (Kleiman *et al.*, 1986). Despite meticulous planning and intensive monitoring subsequently, many of the problems predicted for the orang-utan, above, were observed, such as problems of locomotion through the forest by adults, losses through predation, starvation and disease or for unknown reasons. However, many potential problems of inadequate foraging ability and selection of the highly specialised diet had been anticipated by pre-release training (1.3.6), which proved its value. Re-introduction of this specialist, forest species on a larger scale will require large investments of time and effort.

Animal species with complex life-cycles and/or very specific habitat requirements will be less easy to re-establish. Two butterflies of English fenland habitats could only be re-established for periods of a few years because prevailing drainage practices in surrounding farmland caused adverse changes to the single food plant species of each. The plants were reduced in abundance or suffered a change in growth form (1.2.2). The re-introduction of the large blue butterfly to England has started (Regan, 1986) but many aspects of its habitat and community structure have to be kept within very narrow tolerances, by active management, for any reasonable chance of success. The research effort needed before such a re-introduction may be too daunting in some cases, and the susceptibility of such insect populations to minor habitat changes may often be the cause of apparently successful re-establishments for a few years, to be followed by swift extinction again (see, for example, Duffey, 1977).

Species which migrate regularly between at least two areas will be difficult to re-introduce if no wild conspecifics remain as teachers. The attempt to persuade captive-bred geese to re-occupy vacant, former summer habitat in the Aleutian islands and to migrate to the species' traditional, Californian wintering grounds by imitation of wild-caught geese translocated to the empty habitat has not been successful (1.3.6). When traditional migration patterns are maintained by cultural knowledge, any migration in a re-introduced population, if it evolves at all, will be unlikely to resemble that of the aboriginal population. Migrant animals can have complex life-cycles and important roles in ecosystems dynamics. Some 40 species of sphingid moth in Costa Rica migrate 15–50 km between Pacific coast dry forest and Atlantic coast rainforest, breeding seasonally in each habitat. The migrations of adult moths are largely maintained by cultural tradition, and the populations require recruitment in both habitats (Janzen, 1986). It would be hard to conceive of any moths re-introduced into either or both forests migrating in the present way. This would result in either re-introduction failure or population persistence only at low levels in one or both forests.

The return of some species may be hindered by their real or supposed danger to people. Public opinion would probably not tolerate the return of the wolf or bear to the Scottish highlands, as has been proposed (Yalden, 1986). Re-introduction of the lynx into Austrian forests was preceded by an outcry that they were a threat to people, but very few people subsequently saw the animals, and their impact on the deer population was negligible (Festetics, 1981). Public apprehension is more justified when large animals, such as chimpanzees, lions or tigers, are released when highly habituated to people. These species have all caused death or injury, or are suspected of doing so (see, for example, Adamson, 1986; Brewer, 1978; Singh, 1984). The acceptability of these individual re-introductions, often to re-inforce existing populations, may depend on selection of an appropriate area with least chance of conflict with human interests.

These cases are beyond the mainstream of population re-introductions. In general, specialists and species that are rare because of occupying niches in high trophic levels will require greater effort to re-establish. Unfortunately, in the case of birds, many endangered species have low reproductive rates, specialised ecological requirements and solitary social organisations, all of which makes them poor prospects for successful re-introduction by comparison with birds which have been successfully re-established (Fyfe, 1978). Although the costs for these less re-introducible species will be higher, re-establishment will not be impossible. The most crucial step is to understand the animal and to see how its biology can be manipulated by management techniques, with the aim of overcoming the species' most likely challenges.

10.5   The similarities between aboriginal and re-established populations

It was argued that the Arabian oryx re-established in Oman was probably exploiting its desert habitat differently from the original population, and that this would continue even when the animals' collective knowledge was greater than after the first 5 years back in the wild (9.6). Conditions in the desert changed during the period of extinction, and new traditions of knowledge and use would be expected to develop. How characteristic is this of re-introduced populations?

The European bison, like the Arabian oryx, had been absent from the wild for 33 years when zoo stock were re-introduced into natural habitat in 1952.

The population is still not totally independent, for it receives some winter feeding to reduce mortality and avoid tree damage (Pucek, 1984). Its primaeval forest habitat has also been modified (10.3), and the bison is agreed to differ slightly from its forebears (Krysiak, 1967). Slight differences in behaviour are the beginning of a continuum, for the restored peregrine falcon population of the eastern USA is derived from genetic stocks completely different from those of its predecessor (1.2.4). Patterns of habitat use have changed in that the new population nests on buildings and hack-towers instead of only on cliffs, but ecologically it fills the same niche, and looks the same to most observers (Barclay & Cade, 1983). Straining inclusion as a re-introduction, rather than an introduction, is the example of the whooping crane in the USA. By successful cross-fostering with sandhill cranes on the latter's breeding grounds, whooping cranes now migrate with their foster parents and use the latter's winter and summer ranges (Drewien & Bizeau, 1978). These, however, do not overlap with either the whooping crane's remnant, contemporary range in Canada nor their known historical one. Yet, it is claimed that the USA has a restored, wild population of whooping cranes.

It may be to little advantage to take too puritan an attitude, and to discount as true re-introductions species which show even behavioural differences from their predecessors. The behaviour of wild populations can change suddenly, sometimes with fundamental consequences for their viability. When the endemic Mauritius kestrel had dwindled to six birds, one pair fortuitously nested on a cliff instead of a tree cavity. The young, protected on the cliff from introduced predators, survived and themselves bred on cliff sites. Within four years the population had trebled in size because of this tradition shift (Temple, 1978). Managing bird behaviour to establish new traditions is suggested as a potent technique for adapting behaviour to the exigencies of changed environments. As habitats may change, even through short extinction periods, behavioural modification and consequent niche change may be a small cost for the continuing presence of familiar and established species or populations (Temple, 1978). When a re-introduction is contemplated, it is probably prudent to assume changes in either the habitat or community structure, to which the returning population will have to respond. Moreover, if the extinction period has exceeded 50 years, it is most unlikely that knowledge of the aboriginal population will be detailed enough to allow any meaningful comparison with the restored one.

10.6   Re-introduction costs and values

Oman was aware that restoration of the oryx would be expensive (Daly, 1985), and this still proves to be the case (9.2.2). Much of the costs of any properly designed re-introduction will be made up of salary costs for monitoring.

Re-introduction costs will vary with very obvious factors such as the size and number of animals involved, the size of area they move through, the animals' habits, and the frequency of sighting or detail of information after release that is required. Few re-introductions have been costed precisely, but some translocations also provide sound data.

The real cost of each European eagle owl for its rearing, release and monitoring was DM.1000 (US$444) in 1981, and in that year alone 160 owls were released (von Frankenberg *et al.*, 1983). The two-year captive-breeding programme for the Lord Howe Island woodhen cost Aust.$268000 for the production of 78 birds to re-inforce the remnant population (Fullagar, 1985). Before this intervention, the endemic species had appeared to be doomed. The provisional estimate for re-establishing 12 red wolves is US$135000 in each of 5 years, covering all travel, veterinary and monitoring costs and staff salaries (Parker, 1986).

Translocation of 36 muskox within Canada cost $1000 per animal, exclusive of salaries and monitoring costs (Coady & Hinman, 1984). On the other hand, the very careful capture and movement of 131 baboons in three troops cost US$500 per head, but this included a year's preliminary work to locate destinations and six months of monitoring (Strum & Southwick, 1986). In the baboon case, survival was very high, in marked comparison with that of black-tailed deer in California. Each deer taken from an island in response to overstocking cost US$431 to move. After one year's mortality in the new habitat, the cost per survivor had risen to $2876. Thus, the exercise was neither a cost-effective nor humane solution to the excess numbers (O'Bryan & McCullough, 1985).

In general, re-introductions are expensive conservation projects, largely because they are labour intensive and monitoring must continue over several years. Against this, however, should be set their benefits and values, many of which cannot be easily assessed conventionally. The main benefits are:

(1)   A natural area with as nearly complete a community as possible is intrinsically more satisfying. This has been expressed as the Mountains of the Moon without mountain gorillas seems like saving the husk without the kernel (Conway, 1986). It is a particularly convincing reason for the restoration of large,

conspicuous species such as condors, the oryx and Przewalski's horse.

(2) Efforts to secure the habitat of a re-establishing population protects other species. The oryx project in Oman ensured effective protection for Arabia's largest, wild population of Arabian gazelles (4.6.1). Jamaican hutias were released on to private land, which would then receive extra protection, a significant development in an island with no established system of protected areas (Oliver, 1985). The main purpose served by rehabilitating lions to re-inforce a wild population may have been the attention drawn to the release area and the struggle to prevent its illegal occupation by pastoralists (Adamson, 1986).

(3) A returned species restores ecological processes. This is obvious in the case of scavengers such as the sea eagle, griffon vulture and, perhaps one day, the Californian condor. Ants were watched removing shed hairs from oryx lying sites on the oryx's first day in the Oman enclosure. Thus, the desert's largest wild herbivore was already contributing to ecosystem energetics after a minimum 10 year gap. The returned oryx subsequently diversified the area's ecological linkages in many ways.

(4) The re-introduced species becomes a symbol for habitat pre-servation, and then stimulates environmental awareness and concern in local populations. The golden lion tamarin re-introduction was accompanied by an exhaustive public information and awareness campaign, reaching 89 000 people in the surrounding areas (Kleiman *et al.*, 1986). Surveys and extension work showed the local population's environmental concerns and its under-standing of the benefits derived from native forests, the home of the tamarin (Dietz, 1985).

(5) The return of a species presents unique opportunities for education and training. The tamarin re-introduction project worked with local schools, and satisfied a few of the flood of applicants wishing to work as interns or researchers (Dietz, 1985). The first pink pigeons were released near the Madagascar capital to show them to local people and draw their attention to the plight of the island's endemic species (Todd, 1984), even though the release area was not pristine habitat (1.2.2). After the Cévennes Park authorities had released beavers, a travelling display visited schools and riparian-land owners to explain how people and beavers could live together (Parc National des Cévennes, 1983). The degree of success in

genuinely rehabilitating orang-utans to the wild has led to its value being questioned. There is no doubt, however, of the impact on young Indonesians of visiting combined rehabilitation and education centres to see semi-independent animals and to be exposed to the need for forest and habitat conservation (Aveling & Mitchell, 1980). The red squirrel was re-established in London's Regent Park for several reasons, including those of enabling the public to see again a much-missed species, and to draw attention to the growing use of re-introduction as a conservation management method (Bertram & Moltu, 1986).

(6) Re-introductions offer economic opportunities. The return of the oryx to Oman was partially stimulated by the need to improve the economic prospects of desert tribesmen (3.7.1), in which it succeeded (Chapter 8). Effective enforcement of the protected status of the tamarin's forest depended on the recruitment and training of local people as guards (Kleiman *et al.*, 1986). Without the re-introduction this critical step for maintaining the forest would not have been taken. When there is a good chance of seeing released animals, a re-introduction has tourist potential. In the summer months, griffon vultures in aviaries in the Cévennes and the neighbouring, re-established colony on cliffs have become a major tourist attraction (Fonds d'intervention pour les rapaces, 1985).

Public acceptability of a re-introduction, which is so essential (Fig. 10.1), can be turned into public enthusiasm with little difficulty. Local communities overwhelmingly welcomed the proposed return of the red wolf (Parker, 1986). The Omani public, through newspaper and television coverage, soon regarded its wild oryx as a unique natural resource, and the same pride is seen with the tamarins, griffon vultures and so on. Public attention is often stimulated by the achievements and setbacks of individual, released animals such as Elsa the lioness (Adamson, 1960). The first scimitar-horned oryx calf, born in Tunisia in semi-natural conditions to re-introduced parents, was widely reported (Gordon, 1988). A re-introduced female lynx who was known to have cubs was presumed dead after her radio-collar was found buried. Her death attracted huge attention in France with a great outcry against her killers (Devitt, 1988). One object of the Wisconsin trumpeter swan recovery program is the establishment of a citizens' group for education and fund raising. Half the financial needs for the 10 year project are to come from the private sector (Matteson, 1987).

There is no doubt that re-introduced animals cause strong feelings of pride and involvement, with a sense of vicarious ownership, in local communities of people. As mere introductions or translocations do not create the same attention nor sense of involvement, re-introductions should be seen as a powerful means of mobilising communities in favour of environmental conservation in the ways listed above.

10.7   Captive breeding and re-introductions

Most re-introductions are of species which are absolutely rare or extinct in the wild. Thus, zoos will continue to be the sources of founder stock. As zoos were, in general, set up for other purposes (Cherfas, 1984), are their methods and practices producing the best possible individuals for re-introduction?

There is now a consensus that to ensure the long-term viability of a captive population, management methods should retain 90 % of the founding genetic variability for 200 years (Soulé *et al.*, 1986). As no wild species has yet been kept continuously in captivity for this period, it cannot be said whether such animals, if released into presumed natural habitat, could ever survive. After only six generations in captivity, red grouse showed genetically based changes in gut length because of their artificial diet (Moss, 1972). Experimental breeding of a small rodent suggested that some hereditary control over habitat selection was lost after 12–20 generations in captivity (Wecker, 1970). Selection in captivity can proceed unnoticed towards gradual domestication, requiring strenuous efforts to avoid it through genetic management (Frankham *et al.*, 1986). It may be necessary to institute totally different management strategies for species which might ultimately be re-introduced and those which are destined solely for a permanent captive existence (Foose *et al.*, 1986).

There are a number of actions which could increase the fitness of captive-bred individuals for release into the wild:

(1) Selective pressures could be made more natural. For example, as a low-cost option, polygynous antelopes could be kept in non-exhibit satellite areas, large enough for a more normal social organization to develop (Doherty, 1982). Breeding males would acquire their status by successful competition rather than through whim or the dictates of an egalitarian breeding programme. Equal representation of different males might have to be compromised by production of higher-quality individuals.

(2) The duration of captive breeding should be minimised. The rationale behind taking the last black-footed ferrets from the wild was to increase their numbers rapidly by the best available captive propagation techniques before restoring them to the wild within a few years (Seal, 1988). In the case of the woodhen, *in-situ* semi-captive breeding achieved very successful re-inforcement over two to three years (Fullagar, 1985).

(3) Captive management methods should take advantage of all relevant observations on the performance of zoo stocks after release into the wild. The phenotypic differences in the oryx hyoid shape (7.4.3) suggest that young zoo animals might be better pre-adapted to the wild by grazing during growth, rather than eating cut feeds. Similarly, the experiences of the first released golden lion tamarins will improve the selection and preparation of subsequent immi-grants (Kleiman *et al.*, 1986). Suitable individuals will be identified on their personality profiles. Experience of other species when wild or semi-wild should allow checks on the behavioural competence of animals for re-introduction. Testing would be particularly appro-priate for dominant males, on whom group cohesion might depend, or on breeding females whose example in the wild might teach inexperienced females.

(4) Increasing numbers of captive populations have co-ordinated and co-operative breeding plans, such as the oryx's Species Survival Plan (9.5.1). If such plans can maintain genetic variability through equalising founders' contributions, and avoid inbreeding while at the same time purging any obvious recessive traits, as in the Hawaian goose (Kear & Berger, 1980), a re-introduction project will have the best possible genetic resources for founder selection.

All but one of the oryx imported into Oman came from the USA. Communication between two cultures, on two continents half-way around the world, caused misunderstandings. Captive-breeding of the Hawaiian goose would have been more successful initially if done in or close to Hawaii with a daylength regime similar to the natural habitat (Kear & Berger, 1980). As the veterinary procedures needed before animals travel internationally grow more complex and inclusive, species may find themselves in the embarrassing position of not meeting the health requirements to return to their native country. The oryx's blue-tongue problems in the USA foreshadow this development (5.2.2). The future is likely to see an increasing proportion of animals re-introduced from captive-breeding institutions in their own country or region. The scope for publicity is greatly enhanced when a national zoo is to provide animals for release into its own country.

It is evident that re-introduction projects will rely on animals from conventional zoos for a long time, if not for ever. Zoos now claim for themselves a significant role in animal conservation as natural habitats and their animals are increasingly imperilled around the world. One demonstration of this would be a very positive commitment to assist re-introductions with stock for release, having ensured that while in their care the species retains its wild-type characteristics. If this is not done, many future releases will not be re-introductions but introductions of feral animals, a far less laudable objective.

10.8   Conclusions: re-introductions in perspective

In contemporary conservation few re-introduction projects have yet achieved total success or even final conclusions, but the number of well planned and imaginative attempts around the world is increasing. To reiterate (1.1), the present rate of species' extinctions is increasing. But, its coincidence with changing conservation attitudes and new technologies all lead to the hope that there will be scope for re-introductions in the future. However, the technique will not suit all species. Many of the species which are re-introducible (10.3) are large occupants of open country, which made them particularly susceptible to extermination or extinction by man. Many characteristics of the oryx, such as its conspicuousness and poor running ability (2.4.2 and 2.4.3), predisposed it to efficient overhunting. The various scavenging birds, which are now being brought back, suffered heavily from poisoned carcasses. These feeding habits are now the basis for their successful returns. Thus, re-introduction may be most promising for species easily eliminated by direct persecution. The factors which predispose species to extinction at the hand of man or through his activities show some consistency (Norton, 1986). Combining these principles with the biological analysis used for the Arabian oryx should allow precise identification of species most suited for re-introduction to the wild. At the moment, both Przewalski's horse and the Californian condor stand out as capable of re-establishment.

The less re-introducible species (10.4) tend to be smaller, occurring at low densities because of ecological specialisation or belonging to the higher trophic levels. These species are difficult to exterminate directly, but are more sensitive to habitat damage or alteration, or extinction through ecosystem synergistic effects (Myers, 1987). Their conservation in the wild is more effectively achieved by preventive action, namely protection of their habitat.

The world-wide number of species in this category outnumbers the large, re-introducible species by many orders of magnitude. Thus, re-introduction proposals must be carefully assessed for their chances of success.

Each re-introduction has far more significance than just the return of a single species to the wild (10.6). The return of the Arabian oryx to Oman's deserts was an ambitious venture, but the results and progress described here will, I hope, stimulate and encourage other re-introductions around the world. Each successful re-introduction is a demonstration that man cares about maintaining or enhancing the diversity of the world's fauna and its ecosystems.

APPENDICES

APPENDIX 1
Names of species mentioned in the text

Addax	*Addax nasomaculatus*
Ass, see Wild ass	
Baboon, olive	*Papio cynocephalus anubis*
Badger, honey	*Mellivora capensis pumilio*
Bear, brown	*Ursus arctos*
Beaver, Canadian	*Castor fiber canadensis*
Beaver, European	*Castor fiber fiber*
Bison, European	*Bison bonasus*
Blue, large	*Maculinea avion*
Bustard, great	*Otis tarda*
Bustard, houbara	*Chlamydotis undulata*
Camel	*Camelus dromedarius*
Caracal	*Caracal caracal*
Caribou	*Rangifer tarandus arcticus*
Cat, wild	*Felis silvestris*
Cheetah	*Acinonyx jubatus*
Chimpanzee	*Pan troglodytes*
Clam, giant	*Tridacna gigas*
Condor, Andean	*Vultur gryphus*
Condor, Californian	*Gymnogyps californianus*
Copper, large	*Lycaena dispar batava*
Coyote	*Canis latrans*
Crane, great sandhill	*Grus canadensis tabidu*
Crane, whooping	*Grus americana*
Deer, black-tailed	*Odocoileus hemionus*
Deer, Père David's	*Elaphurus davidianus*
Deer, red	*Cervus elaphus*
Dhub	*Uromastyx microlepis*
Eagle, golden	*Aquila chrysaetos*
Eagle owl, European	*Bubo bubo bubo*

258

Eagle, sea	*Haliaeetus albicilla*
Elephant, African	*Loxodonta africanus*
Falcon, Eastern peregrine	*Falco peregrinus*
Falcon, prairie	*Falco mexicanus*
Ferret, black-footed	*Mustela nigripes*
Finch-lark, black-crowned	*Eremopterix nigriceps*
Fox, Arabian red	*Vulpes vulpes arabica*
Fox, Arctic	*Alopex lagopus*
Fox, Rüppell's sand	*Vulpes rüppelli*
Fox, swift	*Vulpes velox*
Gazelle, Arabian	*Gazella gazella cora*
Gazelle, goitered	*Gazella subgutturosa marica*
Gecko	*Pristurus carteri*
Gecko	*Pristurus minimus*
Gecko	*Hemidactylus turcicus parkeri*
Gemsbok	*Oryx gazella gazella*
Gerbil	*Gerbillus nanus arabicum*
Goat, domestic	*Capra hircus*
Goat, wild	*Capra hircus aegagrus*
Goose, Aleutian Canada	*Branta canadensis leucopareia*
Goose, Hawaiian	*Branta sandvicensis*
Gorilla, mountain	*Gorilla gorilla beringei*
Grouse, red	*Lagopus lagopus scoticus*
Hare, Arabian	*Lepus capensis omanensis*
Orang-utan	*Pongo pygmaeus*
Oryx, Arabian	*Oryx leucoryx*
Oryx, beisa	*Oryx gazella beisa*
Oryx, fringe-eared	*Oryx gazella callotis*
Oryx, scimitar-horned	*Oryx dammah*
Otter, European	*Lutra lutra*
Otter, sea	*Enhydra lutris*
Owl, little	*Athene noctua*
Pigeon, pink	*Columba mayeri*
Polecat	*Mustela putorius*
Rabbit	*Oryctolagus cunniculatus*
Rat	*Rattus rattus*
Raven, brown-necked	*Corvus ruficollis*
Reindeer	*Rangifer tarandus*
Rhinoceros, black	*Diceros bicornis*
Rhinoceros, Indian	*Rhinoceros unicornis*

Rhinoceros, white	*Ceratotherium simum*
Roan	*Hippotragus equinus*
Robin, Chatham Island black	*Petroica traversi*
Sable	*Hippotragus niger*
Sandgrouse, chestnut-bellied	*Pterocles exustus*
Sandgrouse, coronetted	*Pterocles coronatus*
Sandgrouse, Lichtenstein's	*Pterocles lichtensteinii*
Sandgrouse, spotted	*Pterocles senegallus*
Sheep, bighorn	*Ovis canadensis*
Sheep, domestic	*Ovis aries*
Shrike, great grey	*Lanius excubitor*
Skink	*Scincus mitranus*
Skink, Round Island	*Leiolopisma telfairii*
Snake, cat	*Telescopus dhara*
Snake, sand	*Psammophis schokari*
Springbok	*Antidorcas marsupialis*
Squirrel, red	*Sciurus vulgaris*
Stork, European	*Ciconia ciconia ciconia*
Swallowtail	*Papilio machaon britannicus*
Swan, trumpeter	*Cygnus buccinator*
Tamarin, golden lion	*Leontopithecus rosalia*
Thick-knee, spotted	*Burhinus capensis*
Tiger	*Panthera tigris*
Tit, Chatham Island	*Petroica macrocephala*
Turkey, wild	*Meleagris gallopavo*
Viper, carpet	*Echis carinatus*
Viper, horned	*Cerastes cerastes*
Vulture, bearded	*Gypaetus barbatus aureus*
Vulture, griffon	*Gyps fulvus*
Wild ass, Persian	*Equus hemionus onager*
Wild ass, Syrian	*Equus hemionus hemippus*
Wildebeest	*Connochaetes taurinus*
Wolf	*Canis lupus*
Wolf, Arabian	*Canis lupus arabs*
Wolf, red	*Canis rufus*
Woodhen, Lord Howe Island	*Tricholimnas sylvestris*

APPENDIX 2

Veterinary requirements for importation of oryx into Oman
Schedule 9
Schedule of requirements for import licence and health certificate for
imported oryx
All animals must be accompanied by a Veterinary Health Certificate issued
prior to shipment by a duly authorised Veterinary Officer of the country of
origin, stating the following:

(1) All oryx were born in (state country)
(2) Each of the oryx was removed to the port of embarkation from a
herd in which it had remained continuously from its birth or for the
preceding 12 months and which, at the time of removal to the port
of embarkation, was:
(a) officially declared free from tuberculosis
(b) officially declared free from brucellosis
(c) officially declared free from rinderpest
(d) officially declared free from parasitic mange
(e) officially declared free from foot and mouth disease
(f) officially declared free from anthrax
(g) officially declared free from blue tongue;
(h) on premises on which there had been no clinical evidence of
Johne's disease, trichomonas foetus infection or vibric foetus
infection during the preceding 3 years;
(i) on premises which had been free from rabies during the
preceding 12 months;
(3) All the oryx during the 45 days prior to their entry into the
quarantine station were tested as follows:
(a) The single intradermal tuberculin test with negative results.
(b) By blood sampling: Negative to agglutination test for lepto-
spirosis
(c) By blood sampling: Negative to either (i) serum agglutination
test for brucellosis or (ii) complement
fixation test for brucellosis;
(4) Immediately following the commencement of testing, the oryx had
been isolated from all other animals on their premises of origin;
(5) All the oryx have been vaccinated against:
(a) Foot and mouth disease OA and Asia 1
(b) Anthrax

    (c) Clostridial diseases

    (d) Pasteurellosis;

(6) Immediately before movement from the premises of origin to the quarantine premises the oryx for export, and all other animals on the premises of origin, were inspected and found to be healthy and showing no clinical signs or symptoms of infectious or contagious diseases;

(7) I have received a declaration from the responsible breed society that neither the oryx, nor any of the sires or dams, were known to carry any genetic defect;

(8) All the oryx were treated, under veterinary supervision, with a recognised acaricide within 3 days prior to shipment.

Schedule to certificate

Oryx to which this certificate relates:
Identification marks   Date of birth   Sex

Dated: Signed:
                         Authorised Veterinary Officer
                         of the Government

APPENDIX 3

Details of each oryx received or born in Oman between March 1980 and July 1986, with stud number, sex, stud numbers and names of father and mother, dates of birth and death, inbreeding coefficient, place and conditions of birth. Inbreeding coefficients calculated by Dr Georgina Mace

Name	Stud no.	Sex	Father and Mother	Birth (day/month/year)	Death (day/month/year)	Inbr. coeff.	Birth Locn.	Condn.
Kadil	40	m	10 Riyadh 17 Annie	30/03/72	27/02/86	0.000	PHNX	Zoo
Lubtar	135	m	12 Sherman 30 Jane	22/07/77		0.000	SDWAP	Zoo
Farida	148	f	21 Earl 42 Mazna	20/01/78		0.000	SDWAP	Zoo
Jadib	169	m	21 Earl 30 Jane	05/05/78		0.125	SDWAP	Zoo
Salama	172	f	35 Frank 82 Karubi	17/06/78		0.063	SDWAP	Zoo
Mustafan	174	m	35 Frank 74 Tarifa	28/07/78		0.031	SDWAP	Zoo
Nafis	178	m	35 Frank 78 Hasana	21/08/78	02/02/86	0.063	SDWAP	Zoo
Museba	293	m	28 Wayne 46 Aydah	02/03/79	04/09/83	0.125	GPZ	Zoo
Talama	295	f	67 Antar 82 Karubi	17/03/79	01/01/85	0.000	SDWAP	Zoo
Rahaima	297	f	28 Wayne 121 Rosa	22/03/79	29/09/84	0.094	GPZ	Zoo
Sajba	304	f	65 Seif-al-Islam 125 Bakura	19/04/79	28/11/81	0.000	SDWAP	Zoo
Hamid	306	m	67 Antar 74 Tarifa	03/05/79		0.000	SDWAP	Zoo
Malak	309	m	67 Antar 78 Hasana	21/05/79	06/04/80	0.000	SDWAP	Zoo
Hadya	322	f	65 Seif-al-Islam 39 Millie	27/08/79	05/07/86	0.000	SDWAP	Zoo
Zeena	512	f	67 Antar 153 Safi	14/01/82		0.000	SDWAP	Zoo
Mustarayha	523	f	65 Seif-al-Islam 39 Millie	07/02/82		0.000	SDWAP	Zoo

Name	Stud no.	Sex	Father and Mother	Date of Birth (day/month/year)	Date of Death (day/month/year)	Inbr. coeff.	Birth Locn.	Condn.
Hababa	541	f	67 Antar 113 Kalada	22/04/82		0.000	SDWAP	Zoo
Khalifa	592	m	92 Amir 313 Doha	27/01/83		0.000	JRDN	Encl.
Selma	652	f	67 Antar 172 Salama	12/05/80		0.000	Oman	Encl.
Alaga	653	f	293 Museba 297 Rahaima	01/06/81		0.242	Oman	Encl.
Haleema	654	f	169 Jadib 172 Salama	12/03/82	17/08/84	0.117	Oman	Wild
Mundassa	655	f	135 Lubtar 148 Farida	01/06/82		0.063	Oman	Encl.
Migdhaf	656	m	169 Jadib 172 Salama	18/12/82		0.117	Oman	Wild
Bin Talama 1	657	m	169 Jadib 295 Talama	25/01/83	27/01/83	0.086	Oman	Wild
Kateeba	658	f	169 Jadib 652 Selma	04/02/83		0.059	Oman	Encl.
Fadi	659	m	169 Jadib 322 Hadya	18/03/83		0.047	Oman	Wild
Mafrooda	660	f	135 Lubtar 148 Farida	30/04/83		0.063	Oman	Encl.
Sulayman	661	m	135 Lubtar 652 Selma	29/10/83		0.035	Oman	Encl.
Bin Salama	662	m	169 Jadib 172 Salama	09/11/83	06/12/83	0.117	Oman	Wild
Rahman	663	m	169 Jadib 297 Rahaima	09/11/83		0.078	Oman	Wild
Askoora	664	f	169 Jadib 322 Hadya	30/11/83		0.047	Oman	Wild
Bin Talama 2	665	m	169 Jadib 295 Talama	07/12/83	07/12/83	0.086	Oman	Wild
Bint Alaga	666	f	169 Jadib 653 Alaga	29/12/83	29/12/83	0.102	Oman	Wild
Bin Farida 1	—	?	135 Lubtar 148 Farida	30/01/84	30/01/84	0.063	Oman	Encl.
Qurtaasa	720	f	135 Lubtar 652 Selma	31/07/84		0.035	Oman	Wild
Dtheeyab	721	m	306 Hamid 172 Salama	06/08/84		0.047	Oman	Wild

Name	Stud no.	Sex	Father and Mother	Date of Birth (day/month/year)	Date of Death (day/month/year)	Inbr. coeff.	Birth Locn.	Condn.
Tohayla	732	f	135 Lubtar 148 Farida	04/12/84		0.063	Oman	Wild
Jumaa	733	m	306 Hamid 322 Hadya	21/12/84		0.059	Oman	Wild
Samir	742	m	306 Hamid 653 Alaga	03/02/85		0.008	Oman	Wild
London	747	m	135 Lubtar 512 Zeena	14/02/85		0.043	Oman	Wild
Mahyann	765	m	306 Hamid 655 Mundassa	29/03/85		0.012	Oman	Wild
Rasheeda	779	f	135 Lubtar 652 Selma	05/05/85		0.035	Oman	Wild
Murshid	780	m	306 Hamid 172 Salama	07/05/85		0.047	Oman	Wild
Antar	781	m	135 Lubtar 658 Kateeba	19/05/85		0.095	Oman	Wild
Mahmood	*	m	306 Hamid 322 Hadya	18/01/86		0.059	Oman	Wild
Kasheeda	*	f	135 Lubtar 658 Kateeba	23/02/86	17/03/86	0.095	Oman	Wild
Bin Farida 2	*	?	135 Lubtar 148 Farida	01/03/86	05/03/86	0.063	Oman	Wild
Badla	*	f	135 Lubtar 512 Zeena	20/03/86		0.043	Oman	Wild
Bin Alaga	*	m	306 Hamid 653 Alaga	01/04/86	07/04/86	0.008	Oman	Wild
Saleem	*	m	306 Hamid 172 Salama	20/04/86	10/06/86	0.047	Oman	Wild

Locn, location of birth.
Condn, condition of birth.
Inbr. coeff., inbreeding coefficient.
*, Studbook number to be assigned.
?, Sex unknown.
PHNX, Phoenix Zoo.
SDWAP, San Diego Wild Animal Park.
GPZ, Gladys Porter Zoo, Brownesville.
JRDN, Shaumari Reserve, Jordan.
Encl., enclosure at Yalooni.

APPENDIX 4
The program RANGE

RANGE, written by Dr Rudolf Nottrott, is a computer program for processing gridded animal distribution data and for carrying out analyses of area occupancy. It is menu-driven for convenience, and is a two-stage process.

1. Data preparation
   A master database is created in dBase III with the following structure:

Field	Field name	Type	Width	Decimal
1	DAY	Numeric	2	0
2	MONTH	Numeric	2	0
3	YEAR	Numeric	2	0
4	IDENT	Character	3	
5	UTM	Character	6	
6	GRIDSIZE	Numeric	2	0
7	SAFARI	Character	1	
8	COMMENT	Character	20	

Locations (UTM) are entered on a 1 km × 1 km grid scale (GRIDSIZE). Particular subsets of data are then extracted for analysis, e.g. all locations (UTM) of herd 1 (IDENT) between 1 October 1984 and 15 January 1985, using dBASE instructions to define these specifications. The field SAFARI identifies locations on forays or journeys which can be specifically excluded from range area calculations. The three-column field IDENT allows separation of subgroups with their own locations, e.g. Herd 1a, 2f. The COMMENT field can be used, for example, to record the identities of oryx in such subgroups.

2. Data analysis
   When an appropriate subset has been identified, its contents are converted to ASCII format by a dBase instruction. This file is then analysed by RANGE, written in Turbo Pascal (Borland International). Some procedures require two such files for comparison or combination.
   The Main menu offers the following options.

    (1) Change to the Set-up menu (see below).
    (2) A point plot of all grid cells occupied.
    (3) A trace connecting sequentially occupied cells.
    (4) Blocking out of occupied cells, constructing a polygon around them and totalling the enclosed area.

(5) Cumulative range increase from combining two files as in (4) above.
(6) Cells around a trace as in (3) are blocked out.
(7) A frequency distribution of distances between consecutive (daily) locations.
(8) A frequency distribution of distances between two herds on the same date.
(9) A frequency distribution of the distance of a herd from a specified point, e.g. the release site.
(10) Exit to MS-DOS.

The Set-up menu specifies the parameters for the Main menu.

(a) Window size; fixed windows are available for:
Overview, 10 km × 10 km grid.
North window, 2 km × 2 km grid.
Yalooni window, 2 km × 2 km grid.
South window, 2 km × 2 km grid.

User can specify own window location and gridsize.

(b) No. of points per sighting; default 1.
Reference location for (9) above.
Distance and frequency limits for frequency distributions.
Grid on screen: on/off.
Trace stepping, controlled from keyboard: on/off.
(c) Specify input file(s).
Specify disc file for calculation of cumulative range area.
(d) Return to Main Menu.

The results for area occupancy and expansion etc. in Chapter 6 were all obtained from RANGE. With the overview window, a large grid size and the trace step option 'on', a trace allows objective identification of major moves between successive ranges and determination of their dates. Changes in distribution which were forays or journeys are apparent and can be labelled as such (SAFARI). This preliminary examination allows the general database to be divided into sections appropriate for any analysis. Because ranges or range development appear on the screen, processing can be halted at any point for inspection or for printing a map by the Print Screen instruction.

RANGE, therefore, allows an accurate and interactive plot of the areas progressively used by an oryx herd, or the overlap between herds' ranges, their distances apart, or distance from a fixed location. The great bulk of data from daily locations of each herd and subgroup was ideal for RANGE's analytical capabilities.

APPENDIX 5

The size and composition of groups of oryx between 15 June 1984 and 3 August 1984, following the meeting of herds 1 and 2

'$X:Y$' denotes X oryx and Y oryx originally from herds 1 and 2, respectively. Composition is shown only after each change, not necessarily every day.

						Date				

June	15		16		17		18	19	20	21
Groups	a.m.	p.m.	a.m.	p.m.	a.m.	p.m.	a.m.	a.m.	a.m.	a.m.
	9:6	9:10	8:6	0:10	3:1	0:9	11:2	11:2	12:2	12:2
	0:4	3:0	0:5	3:1	2:0	3:0	1:0	1:0	1:9	1:9
	3:0	1:0	2:0	2:0	5:8	8:2	1:9	1:9		
	1:0	0:1	1:0	1:0	1:1	2:0				
	0:1		2:0	2:0	2:0					
				5:0						

June	22		23	24		25		26	
Groups	a.m.	p.m.	a.m.	a.m.	p.m.	a.m.	p.m.	a.m.	p.m.
	13:2	13:1	13:1	13:1	12:1	13:1	13:1	13:3	13:3
	0:9	0:10	0:10	0:10	1:10	0:10	0:9	0:7	0:8
							0:1	1:0	

June	27	28		29		30		July 1		2
Groups	a.m.	a.m.	p.m.	a.m.	p.m.	a.m.	p.m.	a.m.	p.m.	a.m.
	13:2	13:2	13:2	13:7	13:3	11:2	13:1	13:1	13:1	13:1
	0:9	0:9	0:8	0:3	0:8	0:8	0:10	0:9	0:10	0:9
			0:1	0:1		3:0		0:1		0:1

July	2	3		4		5	6	7	8	9
Groups	p.m.	a.m.	p.m.	a.m.	p.m.	a.m.	p.m.	p.m.	a.m.	a.m.
	13:1	13:1	13:1	11:9	12:2	7:1	13:3	13:3	13:4	11:10
	0:10	0:10	0:9	2:0	1:8	6:9	0:7	0:8	0:6	0:1
			0:1		0:1	0:1	0:1		0:1	0:1
										1:0

July	9	10		11	12	13		14		15
Groups	p.m.	a.m.	p.m.	p.m.	a.m.	a.m.	p.m.	a.m.	p.m.	a.m.
	10:1	12:1	13:1	13:1	12:1	12:2	12:1	13:8	13:3	13:3
	2:9	0:9	0:10	0:9	0:9	0:7	0:8	0:1	0:6	0:7
	1:0	1:0		0:1	1:0	0:2	0:2	0:2	0:2	0:1
	0:1	0:1			0:1	1:0	1:0			

July Groups	15 p.m.	16 a.m.	17 a.m.	18 a.m.	20 a.m.	20 p.m.	21 a.m.	23 a.m.	23 p.m.	24 a.m.
	13:3	13:2	9:3	13:3	9:5	8:5	9:5	10:3	9:11	7:11
	0:8	0:8	0:8	0:8	0:6	0:6	0:6	0:7	4:0	5:0
		1:0	3:0		2:0	4:0	4:0	3:0		1:0
			1:0		2:0	1:0		1:0		

July Groups	24 p.m.	25 a.m.	26 a.m.	26 p.m.	27 a.m.	27 p.m.	28 a.m.	28 p.m.	Aug. 3 a.m.
	7:11	12:4	9:4	11:4	11:7	11:7	11:3	11:3	12:3
	6:0	0:5	0:5	0:5	0:2	1:4	1:7	1:8	1:8
		1:2	2:0	1:2	1:2	1:0	0:1	1:0	
			1:2	1:0	1:0		1:0		
			1:0						

REFERENCES

AAZPA (American Association of Zoological Parks and Aquaria) (1987). *Species survival plan: masterplan for Arabian oryx*. Wheeling: AAZPA Conservation office.

Adamson, G. (1986). *My pride and joy: an autobiography*. London: Collins Harvill.

Adamson, J. (1960). *Born free: a lioness of two worlds*. London: Wm. Collins, Sons & Co, Ltd.

Adamson, J. (1969). *The spotted sphinx*. London: Collins & Harvill Press.

Albon, S. D., Mitchell, B., Huby, B. J. & Brown, D. (1986). Fertility in female Red deer (*Cervus elaphus*): the effects of body composition, age and reproductive status. *Journal of Zoology, London, ser*. A, **209**: 447–60.

Amoroso, E. C. & Marshall, F. H. A. (1960). External factors in sexual periodicity. In *Marshall's physiology of reproduction*, ed. A. S. Parkes, pp. 707–831. London: Longman's.

Anderegg, R., Frey, H. & Muller, H. U. (1983). Reintroduction of the bearded vulture or lammergeier, *Gypaetus barbatus*, to the Alps. *International Zoo Year Book*, **23**: 35–41.

Anderson, J. L. (1986). Restoring a wilderness: the reintroduction of wildlife to an African national park. *International Zoo Yearbook*, **24/25**: 192–9.

Anon. (1952). Oryxes caught alive. *Journal, Bombay Natural History Society*, **50**: 186.

Anon. (1973). *The shorter Oxford English dictionary*, 3rd edn, ed. C. T. Onions, p. 1663. Oxford: Clarendon.

Anon. (1979). The story of Oman's oil in pictures. *Petroleum Development (Oman) News*, **124**: 6–7.

Anon. (1980). *The Times atlas of the world*. Comprehensive edition. Times Books, London, in collaboration with John Bartholomew & Son Ltd. Edinburgh: John Bartholomew & Son Ltd.

Anon. (1986). Release of Andean condors. *New Scientist*, **112** (1534): 17.

Aronson, D. R. (1980). Must nomads settle? Some notes towards policy on the future of pastoralism. In *When nomads settle*, ed. P. S. Salzman, pp. 173–4. New York: Praeger Publishers.

Aveling, R. & Mitchell, A. (1980). Is rehabilitating orangutans worth while? *Oryx*, **XVI**: 263–71.

Barclay, J. H. & Cade, T. J. (1983). Restoration of the peregrine falcon in the eastern United States. *Bird Conservation*, **1**: 3–40.

Benirschke, K., Adams, F. D., Black, K. L. & Gluck, L. (1980). Perinatal mortality in zoo animals. In *The comparative pathology of zoo animals*, ed. R. J. Montali & G. Migaki, pp. 471–81. Washington: Smithsonian Institution Press.

Berger, A. J. (1978). Reintroduction of Hawaiian geese. In *Endangered birds: management techniques for preserving threatened species*, ed. S. S. Temple, pp. 339–44. Wisconsin: University of Wisconsin Press; London: Croom Helm Ltd.

Bertram, B. C. R. & Moltu, D.-P. (1986). Reintroducing red squirrels into Regent's Park. *Mammal Review*, 16(2): 81–8.

Beydoun, Z. R. (1980). Some Holocene geomorphological and sedimentological observations from Oman and their palaeogeological implications. *Journal of Petroleum Geology*, 2: 427–37.

Bloxam, Q. M. C. (1982). The feasibility of reintroduction of captive-bred Round Island skink, *Leiolopisma telfairii* to Gunners' Quoin. *Dodo, Journal of the Jersey Wildlife Preservation Trust*, 19: 37–41.

Booth, V. R., Jones, M. A. & Morris, N. E. (1984). Black and white rhino introductions in north-west Zimbabwe. *Oryx*, 18: 237–40.

Borner, M. (1985). The rehabilitated chimpanzees of Rubondo Island. *Oryx*, 19: 151–154.

Brambell, M. R. (1977). Reintroduction. *International Zoo Yearbook*, 17: 112–16.

Brewer, S. (1978). *The forest dwellers*. London: Collins.

Brown, K. (1986). *Prosopis cineraria*. In *Biological resources of significance to people*, Royal Geographical Society Expedition to Wahibah Sands, Sultanate of Oman, Rapid Assessment Document, 1986. London: Royal Geographical Society.

Bruning, D. (1983). Breeding condors in captivity for release into the wild. *Zoo Biology*, 2: 245–52.

Cade, T. J. & Hardaswick, V. J. (1985). Summary of peregrine falcon production and re-introduction by the Peregrine Fund in the United States, 1973–1984. *Aviculture Magazine*, 91: 79–92.

Carruthers, D. (1935). *Arabian adventure to the great Nafud in quest of oryx*. London: Willoughby.

Chatty, D. (1984). *Women's component in pastoral community assistance and development: study of the needs and problems of the Harasis population. Project findings and recommendations*. TCD/OMA-80-WO1/1. New York: United Nations.

Chatty, D. (1986). *From camel to truck: the bedouin in the modern world*. New York: Vantage Press Inc.

Cherfas, J. (1984). *Zoo 2000: a look beyond the bars*. London: BBC Publications.

Clarke, J. E. (1978). *Shaumari wildlife reserve management plan*. Jordan, Amman: Royal Society for Conservation of Nature.

Clutton-Brock, T. H., Guinness, F. E. & Albon, S. D. (1982). *Red deer: behavior and ecology of two sexes*. Chicago: The University of Chicago; Edinburgh: Edinburgh University Press.

Coady, J. W. & Hinman, R. A. (1984). Management of muskoxen in Alaska. In *Proceedings of the first international muskox symposium*, ed. D. R. Klein, R. G. White & S. Keller, pp. 47–51. Biological Papers of the University of Alaska Special Report, no. 4.

Conway, W. G. (1986). The practical difficulties and financial implications of endangered species breeding programmes. *International Zoo Yearbook*, 24/25: 210–19.

Conway, W. G., Foose, T. J. & Wagner, R. O. (1984). *Species survival plan of the American Association of Zoological Parks and Aquaria*. Wheeling, WV: AAZPA.

Council of Europe (1985). Recommendation no. R (85) 15 of the Committee of Ministers on the re-introduction of wildlife species. 4th Meeting.

Daly, R. H. (1985). 'And what', said the Sultan, 'shall we do about the oryx?'. *Species Survival Commission Newsletter*, 5: 25–7.

Dempster, J. P. & Hall, M. L. (1980). An attempt at re-establishing the swallowtail butterfly at Wicken Fen. *Ecological Entomology*, **5**: 327–34.

Devitt, J. (1988). From motherhood to martyrdom. *BBC Wildlife*, **6**(1): 37.

Dieckmann, R. C. (1980). The ecology and breeding biology of the gemsbok *Oryx gazella gazella* (Linnaeus, 1758) in the Hester Malan Nature Reserve. M.Sc. thesis, University of Pretoria.

Dietz, L. A. (1985). Captive-born golden lion tamarins released into the wild: a report from the field. *Primate Conservation*, **6**: 21–7.

Dixon, A. M. (1986). Captive management and the conservation of birds. *International Zoo Yearbook*, **24/25**: 45–9.

Doherty, J. G. (1982). Satellite breeding programmes. Proceedings of AAZPA a Conference: 44–51.

Dolan, J. M. (1976). The Arabian oryx *Oryx leucoryx*: its destruction, captive history and propagation. *International Zoo Year Book*, **16**: 230–9.

Drewien, R. C. & Bizeau, E. G. (1978). Cross-fostering whooping cranes to sandhill crane foster parents. In *Endangered birds: management techniques for preserving threatened species*, ed. S. A. Temple, pp. 201–22. Wisconsin: University of Wisconsin Press; London: Croom Helm Ltd.

Duffey, E. (1977). The re-establishment of the large copper butterfly, *Lycaena dispar batava* Obth., on Woodwalton Fen National Nature Reserve, Cambridgeshire, England, 1969–73. *Biological Conservation*, **12**: 143–58.

Elder, W. H. (1958). *A research report on the Hawaiian goose*. International Committee for Bird Preservation, Pan-American Section, Research Report no. 3: 1–8.

Estes, R. D. (1974). Social organization in the African Bovidae. In *The behaviour of ungulates and its relation to management*, ed. V. Geist & F. Walther, pp. 166–205. IUCN Publications new series no. 24. Morges: IUCN.

Festetics, A. (1981). Die wiederansiedlung des luchses am beispiel der Ostalpen. *Natur und Landschaft*, **56**: 120–2.

Field, C. R. (1975). Climate and the food habits of ungulates on Galana Ranch. *East African Wildlife Journal*, **13**: 203–20.

Finch, V. A. & Western, D. (1977). Cattle colors in pastoral herds: natural selection or social preference? *Ecology*, **58**: 1384–92.

Fitter, R. S. R. (1982). Arabian oryx returns to the wild. *Oryx*, **16**: 406–10.

Flack, J. A. D. (1975). The Chatham Island black robin: extinction or survival? *Bulletin of International Council for Bird Preservation*, **12**: 146–50.

Fonds d'intervention pour les rapaces (1985). Réintroduction du vautour fauve dans les Cévennes. *Rapport d'activité*, pp. 1–4.

Foose, T. J. (1983). The relevance of captive populations to the conservation of biotic diversity. In *Genetics and conservation: a reference for managing wild animal and plant populations*, ed. C. M. Schonewald-Cox, S. M. Chambers, B. MacBryde & L. Thomas, pp. 374–401. Menlo Park, CA: Benjamin/Cummings Publishing Company Inc.

Foose, T. J., Lande, R., Flesness, N. R., Rabb, G. & Read, B. (1986). Propagation plans. *Zoo Biology*, **5**: 139–46.

Frankham, R., Hemmer, H., Ryder, O. A., Cothran, E. G., Soulé, M. E., Snyder, M. & Murray, N. D. (1986). Selection in captive populations. *Zoo Biology* **5**: 127–38.

Franklin, I. R. (1980). Evolutionary change in small populations. In *Conservation biology: an evolutionary–ecological perspective*, ed. M. E. Soulé & B. A. Wilcox, pp. 135–49. Sunderland: Sinauer Associates.

Fullagar, P. J. (1985). The woodhens of Lord Howe Island. *Aviculture Magazine*, **91**: 15–30.

Fyfe, R. W. (1978). Reintroducing endangered birds to the wild: a review. In *Endangered birds: management techniques for preserving threatened species,* ed. S. A. Temple, pp. 323–30. Wisconsin: University of Wisconsin Press; London: Croom Helm Ltd.

Galdikas, B. M. F. (1985). Orangutan sociality at Tanjung Puting. *American Journal of Primatology,* **9**: 101–19.

Geist, V. (1971). *Mountain sheep: a study in behavior and evolution.* Wildlife behavior and ecology series, ed. G. B. Schaller. Chicago and London: The University of Chicago Press.

Geist, V. (1975). On the management of mountain sheep: theoretical considerations. In *The wild sheep of modern North America,* Proceedings of the workshop on the management biology of North American wild sheep, University of Montana, Missoula, USA, 18–20 June 1974.

Gillet, H. (1965). L'oryx algazelle et l'addax au Tchad. *Terre et la vie,* **3**: 257–72.

Gilpin, M. E. & Soulé, M. E. (1986). Minimum viable populations: processes of species extinction. In *Conservation biology: the science of scarcity and diversity,* ed. M. E. Soulé, pp. 13–34. Sunderland: Sinauer.

Glennie, K. W. (1977). Outline of the geology of Oman. Memb, ser Soc. Geol. France, **81**: 25–31.

Gordon, I. (1988). Oryx birth brings hope. *BBC Wildlife* **6**(1): 36.

Goudie, A. & Wilkinson, J. C. (1977). *The warm desert environment.* Cambridge: Cambridge University Press.

Grauvogel, C. A. (1984). Muskoxen of northwestern Alaska: transplant success, dispersal, and current status. In *Proceedings of the first international muskox symposium,* ed. D. R. Klein, R. G. White & S. Keller, pp. 57–62. Biological Papers of University of Alaska Special Report, no. 4.

Green, J., Green, R. & Jefferies, D. J. (1984). A radio-tracking survey of otters, *Lutra lutra,* on a Perthshire river system. *Lutra,* **27**: 85–145.

Greig, J. C. (1979). Principles of genetic conservation in relation to wildlife management in Southern Africa. *South African Journal of Wildlife Research,* **9**(3/4): 57–78.

Grimwood, I. R. (1962). Operation Oryx. *Oryx,* **6**: 308–34.

Grimwood, I. R. (1967). Operation Oryx: the three stages of captive breeding. *Oryx,* **9**(2): 110–122.

Hall, H. T. B. (1977). *Diseases and parasites of livestock in the tropics.* Intermediate Tropical Agriculture Series. Harlow, Essex: Longman Group.

Hallander, H. (1976). A proposed and dropped Swedish great bustard project. In *Reintroductions: techniques and ethics,* ed. L. Boitani, pp. 211–22. Rome: World Wildlife Fund.

Hamilton, R. E. A. (1918). The Beatrix or Arabian oryx (*Oryx leucoryx*) in central Arabia. *Journal, Bombay Natural History,* **26**: 283–4.

Hamilton, W. J., Buskirk, R. & Buskirk, W. H. (1977). Intersexual dominance and differential mortality of gemsbok, *Oryx gazella,* at Namib desert waterholes. *Madoqua,* **10**(1): 5–19.

Hannah, A. C. & McGrew, W. C. (1987). Chimpanzees using stones to crack open oil palm nuts in Liberia. *Primates,* **28**(1): 31–46.

Hannah, A. C. & McGrew, W. C. (1989). Rehabilitation of captive chimpanzees. In *Primate responses to environmental change,* ed. H. O. Box. London: Chapman & Hall, in press.

Harland, W. B., Cox, A. V., Llewellyn, P. G., Pickton, C. A. G., Smith, A. G. & Walters, R. A. (eds) (1982). *A geologic time scale,* Cambridge Earth Science Series. Cambridge: Cambridge University Press.

Hearn, J. P. (1986). Artificial manipulation of reproduction: priorities and practicalities in the next decade. *International Zoo Yearbook,* **24/25**: 148–57.

Henderson, D. S. (1974). The Arabian oryx: a desert tragedy. *National Parks and Conservation Magazine*, **48**(5): 15–21.

Herbert, M. (1976). *The reindeer people*. London: Hodder & Stoughton.

Heslinga, J. (1979). The giant clams of Palau. *Hawaiian Shell News*, **XXVII** (10): 1–6.

Hillman, K. (1982). Follow up to the introduction of black rhinos to Pilanesburg Game Reserve. Unpublished report to Bophuthatswana National Parks 1982.

Holloway, C. W. & Jungius, H. (1973). Reintroduction of certain mammal and bird species into the Gran Paradiso National Park. *Zoologischer Anzeiger*, **191**: 1–44.

Homan, W. G. (1988). The establishment of the World Herd. In *Conservation and biology of desert antelopes*, ed. A. M. Dixon & D. M. Jones, pp. 9–13, London: Christopher Helm Ltd.

Houston, D. (1977). The effect of hooded crows on hill sheep farming in Argyll, Scotland. Hooded crow damage to hill sheep. *Journal of Applied Ecology*, **14**, 17–29.

IIED & WRI (International Institute for Environment and Development & World Resources Institute) (1987). *World resources 1987. An assessment of the resource base that supports the global economy*, pp. 7–24. New York: Basic Books Inc.

IUCN (International Union for the Conservation of Nature) (1980). *World conservation strategy: living resource conservation for sustainable development*. Gland: IUCN.

IUCN (1986). *Proposals for a system of nature conservation areas in the Sultanate of Oman – 1986*. Prepared for the Diwan of Royal Court, Sultanate of Oman by IUCN World Conservation Centre. Gland: IUCN.

IUCN (1987). *The IUCN position statement on translocation of living organisms: introductions, re-introductions and re-stocking*. Gland: IUCN.

Jameson, R. J., Kenyon, K. W., Johnson, A. M. & Wight, H. M. (1982). History and status of translocated sea otter populations in North America. *Wildlife Society Bulletin* **10**: 100–7.

Janzen, D. H. (1986). The eternal external threat. In *Conservation biology: the science of scarcity and diversity*, ed. M. E. Soulé, pp. 286–303. Sunderland: Sinauer.

Jarman, P. J. (1974). The social organization of antelope in relation to their ecology. *Behaviour*, **48**: 215–66.

Jefferies, D. J., Wayre, P., Jessop, R. M. & Mitchell-Jones, A. J. (1986). Reinforcing the native otter *Lutra lutra* population in East Anglia: an analysis of the behaviour and range development of the first release group. *Mammal Review*, **16**(2): 65–79.

Jenkins, A. (1965). Once in Britain. I. The beaver, *Animals*, **5**: 515–17.

Johnstone, T. M. (1977). *Harsusi lexicon*. London: Oxford University Press.

Joint Committee for Conservation of British Insects (1986). Insect re-establishment – a code of conservation practice. *Antenna*, **10**(1): 13–18.

Jones, D. M. (1982). Conservation in relation to animal disease in Africa and Asia. *Symposium of Zoological Society of London*, **50**: 271–85.

Jones, M. (1986). The large blue is back where it belongs. *New Scientist*, **112** (1538): 28.

Jungius, H. (1977). Reintroduction of the Arabian oryx into the Jiddat-al-Harasis, Oman: a feasibility study. Unpublished report to IUCN, Gland.

Jungius, H. (1978a). Criteria for the reintroduction of threatened species into parts of their former range. *IUCN Theatened Deer Programme*, pp. 342–351. Gland: IUCN.

Jungius, H. (1978b). The Arabian oryx: its distribution and former habitat in Oman and the possibility of its reintroduction. Unpublished report to IUCN, Gland.

Jungius, H. (1985). The Arabian oryx: its distribution and former habitat in Oman and its reintroduction. *Journal of Oman Studies*, **8**: 49–64.

Jungius, H. & Loudon, A. (1985). Recommendations for the re-introduction of the Père David's deer to China. IUCN and WWF, Gland, pp. 1–21. Unpublished report.

Kear, J. (1977). The problems of breeding endangered species in captivity. *International Zoo Yearbook*, **17**: 5–14.

Kear, J. & Berger, A. J. (1980). *The Hawaiian goose: an experiment in conservation*. Calton: T. & A. D. Poyser Ltd.

King, J. M. (1979). Game domestication for animal production in Kenya: field studies of the body-water turnover of game and livestock. *Journal of Agricultural Science, Cambridge*, **93**: 71–9.

Kingdon, J. (1982). East African mammals. An atlas of evolution, vol. III, part D *Bovids*. London: Academic Press Inc. (London) Ltd.

Kirkwood, J. K., Gaskin, C. D. & Markham, J. (1987). Perinatal mortality and season of birth in captive wild ungulates. *Veterinary Record*, **120**: 386–90.

Kleiman, D. G., Beck, B. B., Dietz, J. M., Dietz, L. A., Ballou, J. D. & Coimbra-Filho, A. F. (1986). Conservation program for the golden lion tamarin: captive research and management, ecological studies, educational strategies, and reintroduction. In *Primates: the road to self-sustaining populations*, ed. K. Benirschke, pp. 959–80. New York: Springer-Verlag.

Knowles, J. M. (1986). Wild and captive populations: triage, contraception and culling. *International Zoo Yearbook*, **24/25**: 206–9.

Konstant, W. R. & Mittermeier, R. A. (1982). Introduction, reintroduction and translocation of neotropical primates: past experiences and future possibilities. *International Zoo Yearbook*, **22**: 69–77.

Krysiak, K. (1967). The history of the European bison in the Bialowieza primeval forest and the results of its protection. *Acta theriologica*, **12**: 223–31.

Lawton, R. M. (1978). *Vegetation reconnaissance survey for the oryx project*. Project report no. 55, Surbiton: Land Resources Development Centre.

Leader-Williams, N. (1988). *Reindeer on South Georgia: the ecology of an introduced population*. Cambridge: Cambridge University Press.

Leakey, R. R. B. & Last, F. T. (1979). Biology and potential of *Prosopis* species in arid environments, with particular reference to *P. cineraria*. *Journal of Arid Environments*, **3**: 9–24.

Ledger, H. P. (1968). Body composition of East African herbivores. *Symposium of the Zoological Society of London*, **21**: 289–310.

Lindgren, E. J. & Utsi, V. (1980). Status of *Rangifer* in Britain. In *Proceedings of the second international reindeer/caribou symposium, Røros, Norway, 1979*, ed. E. Reimers, E. Gaare & S. Skjenneberg, pp. 744–7. Trondheim; Direktorater for vilt og ferskvannsfisk.

Love, J. A. (1983). *The return of the sea eagle*. Cambridge: Cambridge University Press.

Loyd, M. (1965). The Arabian oryx in Muscat and Oman. *East African Wildlife Journal*, **3**: 124–7.

Luthin, C. S., Archibald, G. W., Hartman, L., Mirande, C. M. & Swengel, S. (1986). Captive breeding of endangered cranes, storks, ibises and spoonbills. *International Zoo Yearbook*, **24/25**: 25–39.

Mace, G. M. (1986). *Status of Arabian oryx in captivity. A report for the IUCN Captive Breeding Specialist Group*. Gland: IUCN.

Mace, G. M. (1988). The genetic status of the Arabian oryx and the design of co-operative management programmes. In *Conservation and biology of desert antelopes*, ed. A. M. Dixon & D. M. Jones, pp. 58–74. London: Christopher Helm Ltd.

MacFarland, W. V. (1964). Terrestrial animals in dry heat: ungulates. In *Handbook of physiology*, section 4 *Adaptation to the environment*, ed. D. B. Dill, E. F. Adolph & C. G. Wilber, pp. 509–39. Baltimore: Waverly Press.

MacKinnon, J. R. (1974). The behaviour and ecology of wild orang-utans (*Pongo pygmaeus*). *Animal Behaviour*, **22**: 3–74.

Maddock, L. (1979). The 'migration' and grazing succession. In *Serengeti: dynamics of an ecosystem*, ed. A. R. E. Sinclair & M. Norton-Griffiths, pp. 104–9. Chicago: The University of Chicago Press.

Mathews, G. V. T. (1973). Some problems facing captive breeding and restoration programmes for waterfowl. *International Zoo Yearbook*, **13**: 8–11.

Matteson, S. (1987). Restoring the trumpeter: an ecosystem approach to management. *The Niche, Newsletter of the Wisconsin Bureau of Endangered Resources*, **2**(3): 1–2.

Mayr, E. (1963). *Animal species and evolution*. Cambridge, MA: The Belknap Press of Harvard University Press.

Mensink, M. (1986). Julia: a gorilla with an identity crisis. *New Scientist*, **110** (1513): 68–9.

Mills, M. G. L. & Retief, P. F. (1984). The effect of windmill closure on the movement patterns of ungulates along the Auob riverbed. In *Proceedings of a symposium on the Kalahari ecosystem*, ed. G. de Graaf & D. J. van Rensburg, Koedoe Supplement, pp. 107–18. Pretoria: National Parks Board of Trustees.

Moss, R. (1972). Effects of captivity on gut lengths in red grouse. *Journal of Wildlife Management*, **36**(1): 99–104.

Mustoe, T. W. (1964). Oryx sightings in Oman. Unpublished letter to Fauna Preservation Society.

Myers, N. (1986). Tropical deforestation and a mega-extinction spasm. In *Conservation biology: the science of scarcity and diversity*, ed. M. E. Soulé, pp. 394–409. Sunderland: Sinauer Press.

Myers, N. (1987). The extinction span impending: synergisms at work. *Conservation Biology*, **1**(1): 14–21.

Nature Conservancy Council (1985). *The sea eagle*. Peterborough: Nature Conservancy Council.

Newby, J. E. (1974). *The ecological resources of the Ouadi Rimé–Ouadi Achim Faunal Reserve, Chad*. Report to FAO, Rome. Direction des Parcs Nationaux et Reserves de Faune.

Newby, J. E. (1985). Large mammals. In *Key environments: Sahara desert*, ed. J. L. Cloudsley-Thompson, pp. 277–90. Oxford: IUCN and Pergamon Press.

Newsome, A. E. & Corbett, L. K. (1975). Outbreaks of rodents in semiarid and arid Australia: causes, prevention and evolutionary considerations. In *Rodents in desert environments*, ed. I. Prakash & P. K. Ghosh, pp. 269–75. Hague: Junk.

Nokes, J. (1965). *Desert prize*. Animals, **6**: 467–70.

Norton, B. G. (1986). On the inherent danger of undervaluing species. In *The preservation of species*, ed. B. G. Norton, pp. 110–37. Princeton: Princeton University Press.

O'Brien, S. J., Wildt, D. E., Goldman, D., Merril, C. R. & Bush, M. (1983). The cheetah is depauperate in genetic variation. *Science*, **221**: 459–62.

O'Bryan, M. K. & McCullough, D. R. (1985). Survival of black-tailed deer following relocation in California. *Journal of Wildlife Management*, **49**: 115–19.

Oliver, W. L. R. (1985). The Jamiacan hutia or Indian coney (*Geocapromys brownii*) – a model programme for captive breeding and re-introduction? In *The management of rodents in captivity*, Proceedings of the tenth symposium of the Association of British Wild Animal Keepers, ed. J. Partridge, pp. 35–52. Association of British Wild Animal Keepers.

Osborne, L. (1985). Progress towards the captive-rearing of great bustards. *Bustard Studies*, **2**: 123–30.

Parc National des Cévennes (1981). Reintroduction du vautour fauve (*Gyps fulvus*). Lâcher de Decembre 1981; bilan et étude de comportement. Unpublished report.

Parc National des Cévennes (1983). *La lettre, Parc National des Cévennes*, no. 56.

Parker, D. H. (1985). *The hydrogeology of the Cainozoic aquifers in the PDO concession area, Sultanate of Oman*. Petroleum Development (Oman) LLC, Exploration Department, Oman.

Parker, W. T. (1986). A technical proposal to reestablish the red wolf (*Canis rufus*) on Alligator River National Wildlife Refuge, North Carolina. US Fish and Wildlife Service. Unpublished report.

Pettifer, H. L. (1980). The experimental release of captive-bred cheetah into the natural environment. *Proceedings of the first world furbearer conference*, ed. J. A. Chapman & D. Pursley, pp. 1001–24. Worldwide Furbearer Conferences.

Philby, H. St.J. B. (1933). *The empty quarter*. London: Constable, and New York: Holt.

Pucek, Z. (1984). What to do with the European bison, now saved from extinction? *Acta Zoologica Fennica*, **172**: 187–90.

Ralls, K. & Ballou, J. D. (1982). Inbreeding and infant mortality in primates. International Journal of Primatology, **3**: 491–505.

Ralls, K., Brugger, K. & Ballou, J. (1979). Inbreeding and juvenile mortality in small populations of ungulates. *Science*, **206**: 1101–3.

Ralls, K., Harvey, P. H. & Lyles, A. M. (1986). Inbreeding in natural populations of birds and mammals. In *Conservation biology: the science of scarcity and diversity*, ed. M. E. Soulé, pp. 35–56. Sunderland: Sinauer.

Regan, S. (1986). The large blue comes home. *Butterfly News*, **8**: 8.

Reichholf, J. (1976). The reintroduction of the beaver (*Castor fiber* L.) in Bavaria: some preliminary results. In *Reintroductions: techniques and ethics*, ed. L. Boitani, pp. 49–54. Rome: World Wildlife Fund.

Renaud, J. & Renaud, J. C. (1984). *Centre de réintroduction des cigognes en Alsace*, pp. 1–21. Hunawihr: J. Renaud & J. C. Renaud.

Rijksen, H. D. (1978). *A field study on Sumatran orang utans: ecology, behaviour and conservation*. Mededelingen Landbouwhogeschool Wageningen 78–2. Wageningen: H. Veenman & Zonen B. V.

Rijksen, H. D. (1986). Conservation of orangutans: a status report, 1985. In *Primates: the road to self-sustaining populations*, ed. K. Benirschke, pp. 153–60. New York: Springer-Verlag.

Rodman, P. S. & Mitani, J. C. (1986). Orangutans: sexual dimorphism in a solitary species. In *Primate societies*, ed. B. B. Smuts, D. L. Cheney, R. M. Seyfarth, R. W. Wrangham & T. T. Struhsaker, pp. 146-54. Chicago and London: University of Chicago Press.

Root, A. (1972). Fringe-eared oryx digging for tubers in the Tsavo National Park (East). *East African Wildlife Journal*, **10**: 155–7.

Ross, K. S. (1985). Aspects of the ecology of Sable antelope, *Hippotragus niger roosevelti* in relation to habitat changes in the Shimba Hills, Kenya. Ph.D. thesis, University of Edinburgh.

Ryder, O. A. & Wedermeyer, E. A. (1982). A cooperative breeding programme for the Mongolian wild horse, *Equus przewalskii* in the United States. *Biological Conservation*, **22**: 259–71.

Ryman, N., Baccus, R., Reuterwall, C. & Smith, M. (1981). Effective population size, genetic interval, and potential loss of genetic variability in game species under various hunting regimes. *Oikos*, **26**: 257–66.

Saiz, R. B. (1975). Ecology and behavior of the gemsbok at White Sands missile range, New Mexico. M.Sc. thesis, Colorado State University.

Sale, J. B. (1986). Reintroduction in Indian wildlife management. *The Indian Forester*, **112**: 867–73.

Sale, J. B. & Singh, S. (1987). Reintroduction of greater Indian rhinoceros into Dudhwa National Park. *Oryx*, **21**: 81–4.

Samour, J. H. (1986). Recent advances in artificial breeding techniques in birds and reptiles. *International Zoo Yearbook*, **24/25**: 143–7.

Sandford, S. (1983). *Management of pastoral development in the third world*. London: John Wiley & Sons.

Seal, U.S. (ed.) (1988). *Reproductive biology of black-footed ferrets and small population biology as they relate to conservation*. New Haven: Yale University Press; in press.

Shepard, O. (1930). *The lore of the unicorn*. London: George Allen & Unwin Ltd.

Shepherd, A. (1965). *The flight of the unicorns*. London: Elek Books.

Sinclair, A. R. E. (1983). The adaptations of African ungulates and their effects on community function. In *Ecosystems of the world*, vol. 13 *Tropical savannahs*, ed. F. Bourliere, pp. 401–26. Amsterdam: Elsevier Scientific Publishing Co.

Sinclair, A. R. E. & Duncan, P. (1972). Indices of condition in tropical ruminants. *East African Wildlife Journal*, **10**: 143–50.

Singh, A. (1984). *Tiger! Tiger!* London: Cape.

Slobodkin, L. B. (1986). On the susceptibility of different species to extinction: elementary instructions for owners of a world. In *The preservation of species*, ed. B. G. Norton, pp. 226–42. Princeton: Princeton University Press.

Smith, T. E. (1984). Status of muskoxen in Alaska. In *Proceedings of the first international muskox symposium*, ed. D. R. Klein, R. G. White & S. Keller, pp. 15–18. Biological Papers of the University of Alaska Special Report No. 4.

Snyder, H. A. & Snyder, N. F. R. (1974). Increased mortality of Cooper's hawks accustomed to man. *Condor*, **76**: 215–16.

Soulé, M. E. (ed.) (1986) *Conservation biology: the science of scarcity and diversity*. Sunderland, MA: Sinauer Associates.

Soulé, M. E., Gilpin, M., Conway, W. G. & Foose, J. T. (1986). The millenium ark: how long a voyage, how many staterooms, how many passengers? *Zoo Biology*, **5**: 101–13.

Springer, P. F., Byrd, G. V. & Woolington, D. W. (1978). Reestablishing Aleutian Canada geese. In *Endangered birds: management techniques for preserving threatened species*, ed. S. A. Temple, pp. 331–8. Wisconsin: University of Wisconsin Press; London: Croom Helm Ltd.

Stanley Price, M. R. (1978). The social behavior of domestic oryx. *African Wildlife Leadership Foundation Wildlife News*, **13** (2): 7–11.

Stanley Price, M. R. (1985a) Game domestication for animal production in Kenya: feeding trials with oryx, zebu cattle and sheep under controlled conditions. *Journal of Agricultural Science, Cambridge*, **104**: 367–4.

Stanley Price, M. R. (1985b) Game domestication for animal production in Kenya: the nutritional ecology of oryx, zebu cattle and sheep under free-range conditions. *Journal of Agricultural Science, Cambridge*, **104**: 375–82.

Stanley Price, M. R., al-Harthy, A. bin H. & Whitcombe, R. P. (1988). Fog moisture and its ecological effects in Oman. In *Arid lands: today and tomorrow,* Proceedings of an international research and development conference Tucson, Arizona, 20–25 October, 1985, ed. E. E. Whitehead, C. F. Hutchinson, B. N. Timmermann & R. G. Varady, pp. 69–88. Boulder, CO: Westview Press.

Stewart, D. R. M. (1963). The Arabian oryx (*Oryx leucoryx* Pallas). *East African Wildlife Journal*, **1**: 103–18.

Stewart, D. R. M. (1964). The Arabian oryx (*Oryx leucoryx* Pallas). *East African Wildlife Journal*, **2**: 168–9.

Stromberg, M. R. & Boyce, M. S. (1986). Systematics and conservation of the swift fox, *Vulpes velox*, in North America. *Biological Conservation*, **35**: 97–110.

Strum, S. C. & Southwick, C. H. (1986). Translocation of primates. In *Primates, the road to self-sustaining populations*, ed. K. Benirschke, pp. 949–58. New York: Springer-Verlag.

Suttie, J. M., Goodall, E. D., Pennie, K. & Kay, R. N. B. (1983). Winter food restriction and summer compensation in red deer stags (*Cervus elaphus*). *British Journal of Nutrition*, **50**: 737–47.

Talbot, L. M. (1960). A look at threatened species. *Oryx*, **5**: 240–7.

Taylor, C. R. (1969). The eland and the oryx. *Scientific American*, **220**: 89–95.

Taylor, C. R. & Lyman, C. P. (1972). Heat storage in running antelopes: independence of brain and body temperatures. *American Journal of Physiology*, **222**: 114–17.

Temple, S. A. (1978). Manipulating behavioral patterns of endangered birds: a potential management technique. In *Endangered birds: management techniques for preserving threatened species*, ed. S. A. Temple, pp. 435–3. Wisconsin: University of Wisconsin Press; London: Croom Helm Ltd.

Temple, S. A. (1983). Is reintroduction a realistic goal? In *Proceedings of the Jean Delacour/IFCB Symposium on breeding birds in captivity*. International Foundation for Conservation of Birds, pp. 597–605. London: Croom Helm Ltd.

Templeton, A. R. (1986). Coadaptation and outbreeding depression. In *Conservation biology: the science of scarcity and diversity*, ed. M. E. Soulé, pp. 105–16. Sunderland: Sinauer.

Templeton, A. R. & Read, B. (1984). Factors eliminating inbreeding depression in a captive herd of Speke's gazelle (*Gazella spekei*). *Zoo Biology*, **3**: 177–99.

Thesiger, W. (1959). *Arabian sands*. London: Longmans, Green & Co. Ltd.

Thomas, B. (1932). Arabia felix: *across the empty quarter of Arabia*. London: Jonathan Cape.

Thomas, W. D., Barnes, R., Crotty, M. & Jones, M. (1986). An historical overview of selected rare ruminants in captivity. *International Zoo Yearbook*, **24/25**: 77–99.

Thomas, J. A. (1980). *The extinction of the large blue and the conservation of the black hairstreak butterflies (a contrast of failure and success)*. Institute of Terrestrial Ecology Annual Report.

Thorp, J. L. (1964). The Arabian oryx 1963–64. *Arizona Zoological Society, Special Bulletin* no. 1, 1–14.

Todd, D. M. (1984). The release of pink pigeons *Columba* (*Nesoenas*) *mayeri* at Pamplemousses, Mauritius – a progress report. *Dodo, Journal of the Jersey Wildlife Preservation Trust*, **21**: 43–57.

Turček, F. J. (1951). Effect of introductions on two game populations in Czechoslovakia. *Journal of Wildlife Management*, **15**: 113–14.

Turkowski, F. J. & Mohney, G. C. (1971). History, management and behavior of the Phoenix zoo Arabian oryx herd, 1964–1971. *Arizona Zoological Society, Special Bulletin* no. 2, 1–36.

UNESCO (1979). Map of the world distribution of arid regions. *MAB Technical Notes 7*. Paris: UNESCO.

Uspenski, S. M. (1984). Muskoxen in the USSR: some results of and perspectives on their introduction. In *Proceedings of the first international muskox symposium*, ed. D. R. Klein, R. G. White & S. Keller, pp. 12–14. Biological Papers of University of Alaska Special Report no. 4.

van der Walt, P. T., Retief, P. F., Le Riche, E. A. N., Mills, M. G. L. & De Graaf, G. (1984). Features of habitat selection by larger herbivorous mammals and the ostrich in the southern Kalahari conservation areas. *Supplement to Koedoe*: 119–28.

van Wijngaarden, W. (1985). *Elephants – trees – grass – grazers. Relationships between climate, soils, vegetation and large herbivores in a semi-arid savanna ecosystem* (*Tsavo, Kenya*). ITC Publication no. 4: 159.

von Frankenberg, O., Herrlinger, E. & Bergerhausen, W. (1983). Reintroduction of the European eagle owl *Bubo b. bubo* in the Federal Republic of Germany. *International Zoo Yearbook*, **23**: 95–100.

Wacher, T. J. (1986). The ecology and social organisation of fringe-eared oryx, on the Galana Ranch, Kenya. D. Phil. thesis, University of Oxford.

Wagner, F. H. (1981). Population dynamics. In *Arid land ecosystems: structure, functioning and management*, vol. 2, ed. D. W. Goodall & R. A. Perry, pp. 125–68. Cambridge: Cambridge University Press.

Walther, F. R. (1978). Behavioral observations on oryx antelope (*Oryx beisa*) invading Serengeti National Park, Tanzania. *Journal of Mammalogy*, **59**: 242–60.

Wecker, S. C. (1970). The role of early experience in habitat selection by the prairie deer mouse *Peromyscus maniculatus bairdi*. In *Behavioral ecology*, ed. P. H. Klopfer, pp. 2–41. Belmont, CA: Dickenson Publishing Company Inc.

Wells, S. M. (1988). Snails going extinct at speed. *New Scientist*, **117**: 46–8.

Whateley, A. & Brooks, P. M. (1985). The carnivores of the Hluhluwe and Umfolozi Game Reserves. *Lammergeyer*, **35**: 1–27.

Whitaker, I. (1986). The survival of feral reindeer in northern Scotland. *Archives of Natural History*, **13**: 11–18.

Wilcox, B. A. (1986). Extinction models and conservation. *Trends in Ecology and Evolution*, **1**: 46–8.

Wilkinson, P. F. & Teal, P. N. (1984). The muskox domestication project: an overview and evaluation. In *Proceedings of the first international muskox symposium*, ed. D. R. Klein, R. G. White & S. Keller, pp. 162–6. Biological Papers of the University of Alaska Special Report no. 4.

Williamson, D. T. (1987). Plant underground storage organs as a source of moisture for Kalahari wildlife. *African Journal of Ecology*, **25**: 63–4.

Winfield, C. G. (1970). The effect of stocking intensity at lambing on lamb survival and ewe and lamb behaviour. *Proceedings of the Australian Society of Animal Production*, **8**: 29–32.

Woodford, M. H. (1989). Veterinary implications for the re-introduction of the Arabian oryx into Saudi Arabia. In *Wildlife conservation and development in Saudi Arabia*, ed. A. H. Abuzinada, P. D. Goriup & I. A. Nader. Riyadh: National Commission for Wildlife Conservation and Development, in press.

Woodford, M. H., Kock, R. A., Daly, R. H., Stanley Price, M. R., Kidner, J., Usher-Smith, J. H. & Emanuelson, K. A. (1988). Chemical immobilisation of Arabian oryx. In *Conservation and biology of desert antelopes*, ed. A. M. Dixon & D. M. Jones, pp. 90–102. London: Christopher Helm Ltd.

Woolley, C. W. (1962). [Letter to New Scientist.] *New Scientist*, **300**: 368–9.

Wotschikowsky, U. (1981). Der luchs – aus der theorie und praxis der wiedereinbürgerung. *Natur und Landschaft*, **56**: 122–4.

Yalden, D. W. (1986). Opportunities for reintroducing British mammals. *Mammal Review*, **16**: 53–63.

Zurowski, W. (1979). Preliminary results of European beaver reintroduction in the tributary streams of the Vistula river. *Acta Theriologica*, **24**(7): 85–91.

INDEX

Page numbers in *italics* refer to figures and tables.